Boundary-Scan Test

Boundary-Scan Test

A Practical Approach

by

Harry Bleeker

Peter van den Eijnden

*Fluke/Philips Test & Measurement,
Eindhoven, The Netherlands*

and

Frans de Jong

*Philips Research Laboratories,
Eindhoven, The Netherlands*

SPRINGER SCIENCE+BUSINESS MEDIA, B.V.

ISBN 978-1-4613-6371-2   ISBN 978-1-4615-3132-6 (eBook)
DOI 10.1007/978-1-4615-3132-6

This printing is a digital duplication of the original edition.

Printed on acid-free paper

CONTENTS

Chapter 3

Chapter 4

Chapter 5

Chapter 6

LIST OF FIGURES

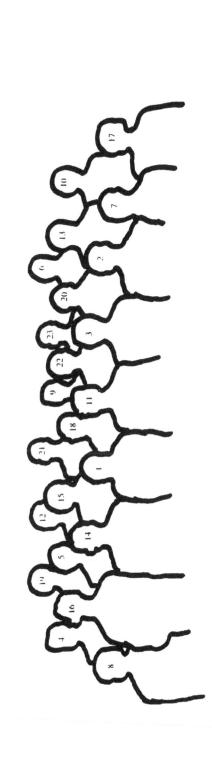

1   Rod Tulloss (AT&T, U.S.A.)
2   Ben Bennetts (Bennetts Associates, U.K.)
3   Colin Maunder (British Telecom, U.K.)
4   Ian Fisher (Computer Automation, U.K.)
5   David Richards (Digital Equipment Corp., U.S.A.)
6   Eskild Jensen (ElectronikCentralen, Denmark)
7   Walter Ghisler (Ericsson, Sweden)
8   Prabhat Varma (GEC HRC, U.K.)
9   Frans Mosselveld (Factron-Schlumberger, The Netherlands)
10   Malcolm Wallace (GenRad, U.K.)
11   Tom Williams (IBM Boulder, U.S.A.)
12   Gordon Hannah (Marconi Instruments, U.K.)

13   Ulrich Ludemann (Nixdorf Computer, Germany)
14   Harry Bleeker (Philips (chairman), The Netherlands)
15   Dirk van de Lagemaat (Philips (secretary), The Netherlands)
16   Frans Beenker (Philips, The Netherlands)
17   Wim Sauerwald (Philips, The Netherlands)
18   Pat Diamond (Plessey (secretary), U.K.)
19   Erwin Trischler (Siemens, Germany)
20   Derek Roskell (Texas Instruments, U.K.)
21   Pete Fleming (Texas Instruments, U.S.A.)
22   Lee Whetsel (Texas Instruments, U.S.A.)
23   Michel Parot (Thomson-CSF, France)

PREFACE

The ever-increasing miniaturization of digital electronic components is hampering the conventional testing of Printed Circuit Boards (PCBs) by means of bed-of-nails fixtures. Basically this is caused by the very high scale of integration of ICs, through which packages with hundreds of pins at very small pitches of down to a fraction of a millimetre, have become available. As a consequence the trace distances between the copper tracks on a printed circuit board came down to the same value. Not only the required small physical dimensions of the test nails have made conventional testing unfeasible, but also the complexity to provide test signals for the many hundreds of test nails has grown out of limits.

Therefore a new board test methodology had to be invented. Following the evolution in the IC test technology, Boundary-Scan testing has become the new approach to PCB testing. By taking precautions in the design of the IC (design for testability), testing on PCB level can be simplified to a great extent. This condition has been essential for the success of the introduction of Boundary-Scan Test (BST) at board level.

Nowadays Boundary-Scan testing is embraced worldwide by almost all PCB manufacturers. The main reason is, of course, cost saving, emerging in various phases of the PCB life cycle. In the *design* phase, testing time is saved during prototyping. In the *factory*, lower test preparation time, time savings in fault diagnosis and much cheaper test equipment has led to a reduction in test costs of over 50%. Moreover, the factory throughput has increased. In the *field service* phase, tremendous savings are made: in the prices of test equipment, in test preparation times and reduced spare board inventories. In this way BST has led to a cost reduction in PCB production of as much as 70%, *after* including the extra development costs for the IC precautions.

The simplicity of BST also makes such cost reductions profitable for small volume PCB production. And since Boundary-Scan Testing involves the whole PCB life cycle, it is more than just the introduction of a new design technology, it is an integral production approach. Therefore BST concerns the global strategy of a company to which also top management should pay attention.

Due to its importance, the architecture of the BST technology has lead to a world standard: IEEE Std 1149.1 [1]. Applying this standard in the IC designs allows PCB designers to use ICs of various manufacturers and yet prepare *standard* test methods and test equipment for the whole life cycle of the product.

This book is an aid to introducing BST into a company. Chapter 1 states the PCB test problem and its solution. Chapter 2 provides a tutorial on the standard for the Boundary-Scan Architecture. The next chapters 3 and 4 describe the innovations for the hardware and the software BST support respectively. Developments in BST are explored with the aid of various examples and results as they have been obtained and published by world leading electronics companies. Chapter 5 describes various basics supporting the actual test technology; which PCB production faults are met, how are they detected and to what degree can they be diagnosed. Chapter 6 gives a management overview of the aspects to be considered in introducing BST in a production organisation, including organizational and cost saving aspects. The book ends with a glossary of terms and an alphabetical index, for quick reference.

The readers of this book are expected to have a basic understanding of ICs and digital circuitry.

Acknowledgements

Many contributions are necessary to complete a book at such a technological level as this. The authors wish to thank their colleagues for their discussions and positive comments in preparing the text. In particular Harm Tiersma should be mentioned here for his remarks concerning the principles of testing and his conscientious reading and correcting of the draft. Special credit must be given to Rien van Erk for his editing and writing efforts, without which this book would probably not have come to fruition.

Chapter 1

PCB TESTING

MINIATURIZATION IN ELECTRONICS

Printed Circuit Boards (PCBs) add the most value to electronics hardware. Over the years, PCBs have become loaded with more components and hence have become increasingly complex and expensive. This is mainly caused by the ongoing miniaturization in electronics.

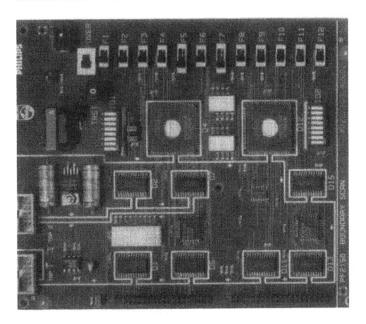

Fig. 1-1   A printed circuit board containing ASICs

To start with, the miniaturization within integrated circuits (ICs) themselves has progressed into the sub-micron technology. This has led to an increased number of gates and a very large number of functions per chip. Consequently, many more pins per IC are needed and hence ICs come in bigger packages with more leads (>500 pins) at smaller pitches (0.3 mm). Therefore, the distance between the foot prints of the ICs on the PCB is becoming equally small.

Direct mounting of chips on both sides of the PCB, using Tape Automated Bonding
(TAB), Chip On Board (COB) etc. have caused further miniaturization of the
hardware implementation.

To even further increase the number of electronic functions per unit of surface,
other technologies are becoming available: Multi Chip Modules (MCMs). These are
functional modules built up of dies directly mounted on top of several dielectric and
metallization layers supported by a ceramic multi-layer substrate. The component
density is very high and accessibility other than via the input and output pins of the
MCM is non-existent.

Figure 1-2 shows the reduction of trace distances on PCBs and in ICs.

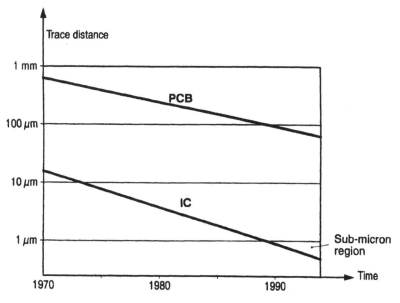

Fig. 1-2   Reduction of trace distances on PCBs and in ICs

Figures 1-3 and 1-4 show, respectively, a part of a PCB and a part of an IC which
is soldered onto that PCB. These pictures are taken with a Scanning Electron
Microscope and the pitch of the IC's contact pins is 25 mil (= $\frac{1}{40}$ inch $\approx$ 0.63 mm).

The ongoing miniaturization has made it more and more difficult to access these
"highly loaded" PCBs mechanically, with fixtures, for in-circuit testing. Moreover,
the test equipment became so expensive in the 1980s that it severely affects the
profitability of producing PCBs. Electronic companies were, therefore, looking for
low-cost test methods based on the sofar hardly explored design for testability.

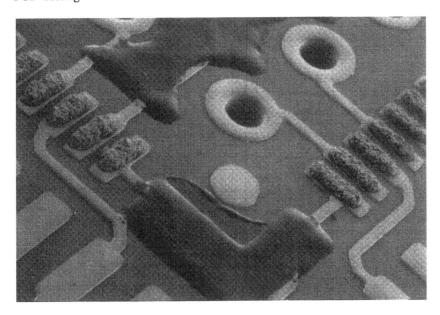

Fig. 1-3 Electron microscope picture of part of a PCB

Fig. 1-4 Electron microscope picture of a soldered IC on a PCB

Figure 1-5 shows a drawing of some different test pins used in a bed-of-nails fixture. The pitch of the test pins is $^1/_4$" and they have a diameter of about 0.8 mm.

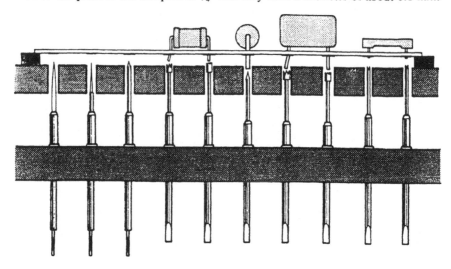

Fig. 1-5   Conventional test pins for a bed-of-nails fixture

It is clear that these pins can not be used for testing purposes at the IC connections in figure 1-4, where the pitch was 0.63 mm.

ROAD BLOCKS FOR CONVENTIONAL PCB TEST METHODS

With *in-circuit testing*, physical contact is made to the components (for example ICs) and the copper tracks on the PCB (test pads) by means of probes or bed-of-nails technology. Input signals are applied to the component's input pins via probes or needles, and the response at the output pins is examined through the probes or needles. For the plated through hole (PTH) technology this method was easy to perform. But the continual miniaturization created the first obstacle to in-circuit testing.

The PCB technology with surface mounted devices (SMDs) *on both sides* of the PCB is another reason why mechanical probing becomes practically impossible.

Next, the advantages of in-circuit testing (ICT), the use of standard test pattern sets for ICs, from which a test program could easily be assembled, have evaporated due to the increasing use of Application Specific ICs (ASICs). Each ASIC requires a separate test set which is not contained in the library of standard test sets. The vendor of the in-circuit test library is unable to supply these tests. Test sets for complex standard VLSI devices (e.g. microprocessors) are not available or very expensive.

Finally, VLSIs require long test sequences, for instance, for initialization. If outputs of ICs are forced (back-driven) during such a sequence, there may be an impact on the overall reliability of the PCBs.

Summarizing, the programming of automatic test equipment (ATE), the mechanical test access and the quality of testing in the conventional way are reaching their limits of feasibility.

The solution to these problems arises from the IC test technology. To overcome the test preparation and fault coverage problems faced in the IC-test technology in the 70's, the scan test technology evolved and became a good solution for digital electronics. The enormous increase in observability and controllability achieved by scan technology forms the base for Boundary-Scan Test technology for PCBs.

THE SOLUTION FOR PCB TESTING

The main point to ensure while manufacturing a PCB is to make sure that all components are mounted properly at the right place on the board and that the interconnections between the components are as prescribed in the design. So, after production, one of the main items to be tested is whether or not the connections between the components (for example ICs) are 100% correct.

The basic idea to circumvent the above mentioned access problem for test purposes was to add a shift register cell next to each input/output (I/O) pin of the component. During test mode these cells are used to control the status of an output pin (high or low) and read the states of an input pin (high or low). This allows for testing the the board interconnections. During normal mode the cells are 'transparent'.

Figures 1-6 and 1-7 show the idea of adding the shift register cells. Figure 1-6 depicts three ICs with 'many' interconnections. The input signals (I) and the output signals (O) are the I/O signals present at the board edge connector.

In figure 1-7 shift register cells have been added between the IC's 'core logic' and the I/O pins to provide a serial test data path through all ICs.

Since the shift register cells are located at the IC's boundary (I/O pins) these cells are referred to as Boundary-Scan Cells (BSCs) and the composed shift register is called the Boundary-Scan Register (BSR). According to the IEEE Std 1149.1 [1], the input to the serial test data path is called Test Data Input (TDI) and the output is called Test Data Output (TDO).

To permit testing of the PCB's connecting copper tracks, the interconnections, the cells must support the following test actions.

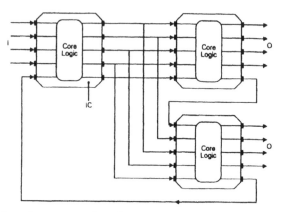

Fig. 1-6   Many interconnections between three ICs

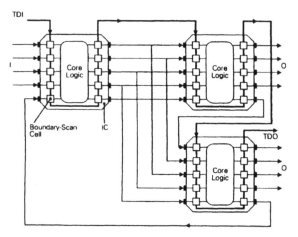

Fig. 1-7   Added shift registers provide scan path

Action 1.   *Shift-DR*: shift stimulus data in from TDI through the registers to the cells related with the output pins of the IC(s).

Action 2.   *Update-DR*: update these output cells and apply the stimuli to the board interconnections.

Action 3.   *Capture-DR*: capture the status of the board's interconnections at the input pins of the receiving IC(s).

Action 4.   *Shift-DR*: shift out the results through the BSR towards TDO for examination.

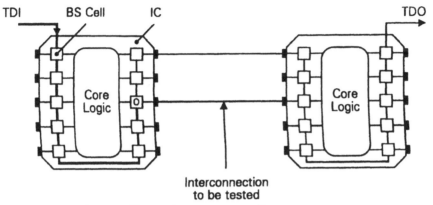

Fig. 1-8   Action 1: shift test data to output cell

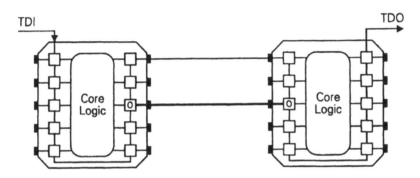

Fig. 1-9   Actions 2+3: apply test data and capture result

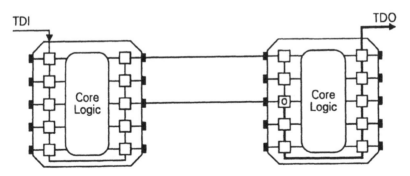

Fig. 1-10   Action 4: shift captured results out

Figures 1-8, 1-9 and 1-10 depict the respective test actions. The test stimulus is indicated here as a bit of value '0' in the Boundary-Scan path.

Note that the actions 2 and 3 perform the actual test: apply test data to the PCB tracks between the ICs and capture the status of these tracks. The two shift actions (1 and 4) do not in fact contribute to the test results.

Testing of the interconnection between the ICs is referred to as external test or *EXTEST*. The circuit diagram in figure 1-11 supports these functions.

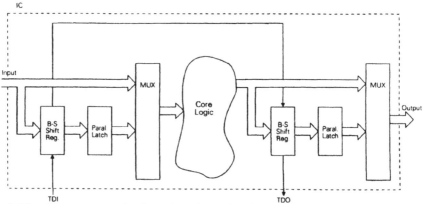

Fig. 1-11   Components of a Boundary-Scan Cell in an IC

Figure 1-12 shows the data flow during the first test action, shift the stimuli (data) through the registers, from input TDI to the output TDO. This action is called Shift-DR. The emboldened lines between TDI and TDO plus the shaded Boundary-Scan (BS) registers depict the signal path.

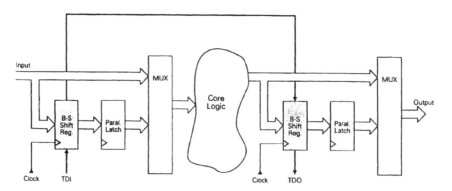

Fig. 1-12   ShiftDR state

Notice that in figure 1-12 the Boundary-Scan design comprises, a parallel latch (flip-flop) and a multiplexer (MUX) in addition to the shift-register stages. The latches have been included to assure a stable signal at the parallel outputs while data (test stimuli and test results) are shifted through the scan path (BSR). The purpose of the multiplexer is discussed later.

The figures 1-13 and 1-14 show the signal flows for Update-DR (above Action 2) and Capture-DR (above Action 3) respectively.

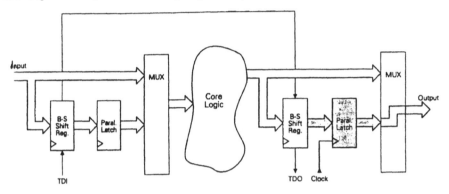

Fig. 1-13   Update-DR state for EXTEST

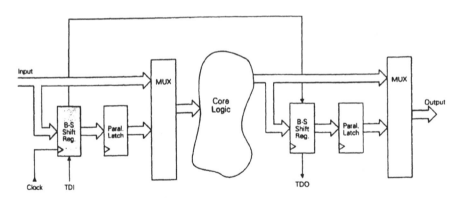

Fig. 1-14   Capture-DR state for EXTEST

Remember that these test steps still concern the EXTEST. The shift state for Action 4 is the same as for Step 1 of course.

The actions indicated in these figures are triggered by the indicated Clock signals. The clock signal triggers the actions in the Boundary-Scan cells. In the Update-DR stage of the EXTEST an electrical signal is then applied to a copper track on the

PCB, which arrives virtually at the same time at the receiving input of the appropriate IC(s), see figure 1-9. Here the Capture-DR action is also triggered by the clock pulse. It is prescribed that the Update-DR action must be triggered at the falling edge (negative slope) of the clock pulse whilst the Capture-DR action is triggered at the rising edge (positive slope) of the clock pulse. Note in the last three figures (1-12 through 1-14) that the EXTEST does not in any way interfere with the functioning of the core logic.

In *conclusion* it can be stated that a board interconnect test can be performed with the aid of Boundary-Scan Registers using the described procedure. It has also been shown that this can be done without interfering with the core logic's operation. Given the registers at the I/O pins of the ICs, it can be seen that an internal test of the IC logic (core logic) can also easily be realized. The testing of the internal logic of an IC is referred to as internal test or *INTEST*. For the INTEST the same test actions as described for the EXTEST are applicable, only the Update-DR and Capture-DR actions concern now the Boundary-Scan cells at the input and output pins of the IC respectively. The following test sequence describes the actions.

Action 1.   *Shift-DR*: shift the data (test stimuli) through the registers to the cells associated with the input pins of the IC(s).

Action 2.   *Update-DR*: update these input cells and apply the stimuli to the core logic.

Action 3.   *Capture-DR*: capture the status of the core logic outputs in the output cells of the IC(s).

Action 4.   *Shift-DR*: shift out the results through the BSR of the IC(s) for examination.

The signal flow during the shift actions is the same as in figure 1-12. The signal flows for the test actions 2 and 3 are shown in figures 1-15 and 1-16 (shaded).

Test Control Logic

As discussed above, several test control signals are used to control the Boundary-Scan shift-registers, the parallel latches and the multiplexers. It would be unacceptable to define an IC package pin for each control signal. Therefore, a control logic has been designed consisting of a sequencer that can be controlled by a serial protocol. In addition to this sequencer an Instruction Register (IR) is required to select the different tests to be performed, for example INTEST and EXTEST. The combinations of the sequencer states and the IR instructions are used to generate the internal control signals for the various test operations.

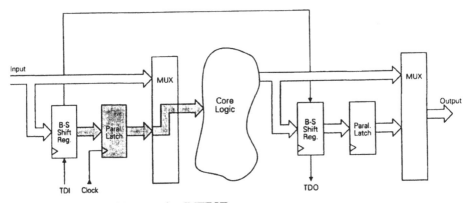

Fig. 1-15   Update-DR state for INTEST

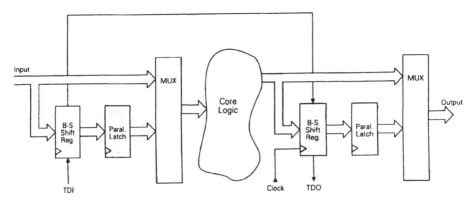

Fig. 1-16   Capture-DR state for INTEST

The sequencer is controlled by two signal lines: Test Mode Select (TMS) and Test Clock (TCK). The whole Boundary-Scan test logic requires at least four dedicated test pins at the IC package, thus additional to its other functional pins. The four pins (TMS, TCK, TDI and TDO) are referred to as the Test Access Port or TAP. The entire test control logic is consequently called the TAP Controller.

Remark.   Since all these test provisions have been standardized in the IEEE Std 1149.1 [1], the 'standard' names for the signals and the IC pins are used further in this book. An overview of the birth of the Boundary-Scan standard is given in chapter 6.

At board level, refer back to figure 1-7, the test control lines TMS and TCK have to be applied to all components on the board that are suited for BST. Figure 1-17 includes the infrastructure for the TMS and TCK signals.

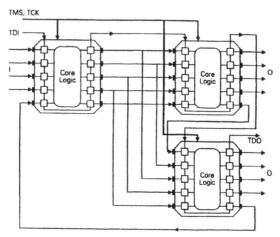

Fig. 1-17   BST infrastructure on PCB

Figure 1-17 shows that the TMS and TCK signals are applied to all relevant
components in parallel (thus all TAP Controllers are always in the same state),
whereas the test path, TDI to TDO, is a serial path all over the printed circuit board.

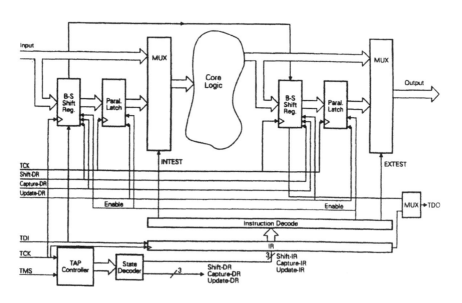

Fig. 1-18   Test control logic for BST

The way that these signals control the Boundary-Scan Test logic is depicted in
figure 1-18. Here it can be seen that, under control of the State Decoder, either the

Instruction Register's serial path or the Boundary-Scan shift register stages can be connected between the IC's TDI and TDO pins. Figure 1-19 shows at IC level the test control logic and the serially connected Boundary-Scan cells. In practice also the IC input and output buffers are present, as drawn in figure 1-19.

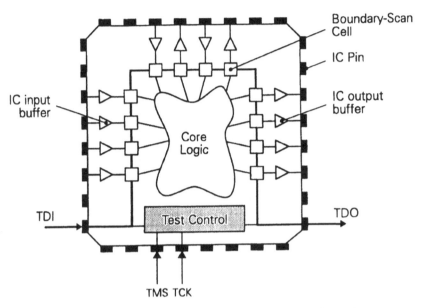

Fig. 1-19   IC including Boundary-Scan Register and Test Control

Figure 1-20 shows an example of a board level Boundary-Scan test configuration consisting of one serial test path.

Remember that the actual testing of the printed circuit board concerns the connections between the ICs, denoted as net*1*, net2 ... net*n* (EXTEST), but the test architecture can also be used for checking the internal operation of the chip's core logic (INTEST).

Example Boundary-Scan Circuits

For a board level test procedure the Boundary-Scan registers should have three important properties:

1. they must be 'invisible' during normal operation of the IC's core logic,

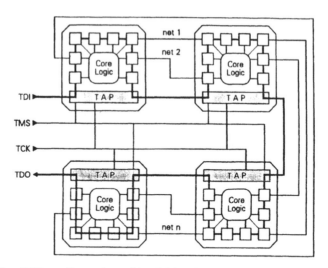

Fig. 1-20   PCB configuration with all IC BSRs connected in series

2. they must be able to isolate the IC interconnections on the board from the IC cores to allow testing the external IC connections, and

3. they should be able to isolate the IC from its surroundings on the PCB to also allow testing of the internal core logic.

An example of a commonly used cell design is given in figure 1-21.

The test properties are obtained as follows. To support normal operation (Test/Normal) the multiplexer 2 (MUX2) is kept at logical '0'. The control signal (Shift/Load) of MUX1 can be kept at '1' or '0'.

The BS cell depicted in figure 1-21 can be used both at the input and at the output related IC pins. For example, for an INTEST of the core logic (see figure 1-19) the BS cells at the IC *input* receive data shifted-in via TDI with MUX1 set to '1' (Shift) and clocked to the core logic with MUX2 set to '1' (Test). At the core *output* the BS cells must capture the resulting output signal for which MUX1 is set to '0' (Data from system). These signals are clocked towards TDO for examination, for which MUX2 is kept at '1.

Note that the above depicted Boundary-Scan cell is just one sample representation. Other circuits and test features are discussed in chapter 2.

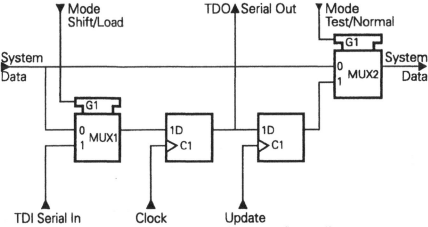

Fig. 1-21   Example of a commonly used Boundary-Scan cell

During testing, an IC's Boundary-Scan Register can also stay idle, thus without interfering with the in-core logic. For example, when the IC is not used in a board test it is senseless to include the IC's BS Register (BSR) in the PCB Boundary-Scan path. In such cases the BSR is short cut by the so called Bypass register, a one-bit register between the TDI and TDO pins to make the board test faster, thus more efficient.

The Bypass Register may be selected by loading the appropriate instruction into the Instruction Register. The presence of instructions opens a wide variety of test possibilities. For instance it is possible to have a Device Identification Register implemented in the component (IC), by means of which it can be tested whether the right component is mounted on the right place on the PCB. Or, it is possible to have manufacturer specific instructions for several purposes, e.g. emulation, self-test, test access to internal scan registers, etc. Figure 1-22 shows an example of an Instruction Register cell.

The IR operation and applications are detailed in chapter 2.

The TAP Controller (see figure 1-18) is a 16-state sequential machine operating synchronously with the TCK and responding to the TMS signal. Under control of the TAP Controller the instruction and test data are shifted from the TDI, through the registers and out again towards the TDO pin.

A Boundary-Scan test consists of four steps:

1. shift in and decode instructions, i.e. select a particular data register,

2. shift in test data,

3. execute the test and

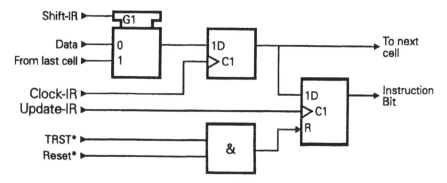

Fig. 1-22   An example of an Instruction Register cell

All the actions can be controlled by the TAP Controller and are described in chapter 2.

Constituent Elements of the BST Path

As mentioned above, applying Boundary-Scan Test (BST) on a PCB implies that Boundary-Scan (BS) cells are placed along the IC's input/output pins. For test purposes, these BS cells replace the conventional *physical* test pins (which can be placed on the IC contact pins during testing) by 'virtual' test pins inside the IC.

In the conventional way an interconnecting copper track (or 'net') between the output of IC1 and the input of IC2 is tested by placing two test pins on the net nodes and applying test signals to one node while measuring the responses at the other node. These nodes (measuring points) could be physically located at the PCB soldering joints of the IC pins.

In the case of BST, the conventional nodes become 'virtual' nodes located inside the BS cells. Including part of the IC itself in the test path means that the test path becomes longer, so more possible connection points are to be considered when a fault is detected. Figure 1-23 (introduced in [2]) gives an idea of the whole BST test path between two ICs.

The signal test path between two 'virtual' test pins now includes:

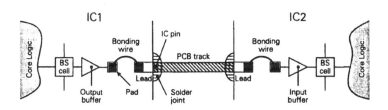

Fig. 1-23 BST path between two ICs

- the Output and Input Boundary-Scan cells,
- the Output and Input buffers,
- the Bonding pads, four in total,
- the Bonding wires, two in total,
- the IC pin soldering joints, two in total,
- the PCB copper track.

Obviously all these points can give rise to faults. Although the ICs are tested usually before leaving the supplier's factory, the ICs can, once mounted on the PCB, show faults. This can be caused by component handling during loading the PCB, for example due to mechanical or thermal shocks.

It should be noted that a deduced 'open' fault (path interrupted) from a BS test does not indicate on which of the above mentioned points the interruption occurs. For example the connections of the two soldering points and the PCB copper track in between can be faultless while the input buffer of IC2 is blown up due to electro static discharge, which then may cause the 'open' error. If one output is connected to several input pins, then further location of such a fault can be deduced with proper test patterns for all input pins involved (for details see chapter 5).

Chapter 2

THE BOUNDARY-SCAN TEST STANDARD

This chapter describes the elements defined by the IEEE Std 1149.1 for the Standard Test Access Port and Boundary-Scan Architecture. First the constituent parts of the Boundary-Scan architecture (both the mandatory and the optional items), are described followed by the instructions. The final section gives some rules which should be followed when BST designs are to be documented. Numerous descriptions are clarified, each with a practical example, making this chapter a tutorial for BST. For more detailed specifications of the BST standard please refer to the listing in the Appendix of this book, where these specifications are summarized for those readers who do not have the official IEEE Std 1149.1 manual readily available.

THE BST ARCHITECTURE

The IEEE Std 1149.1 distinguishes four basic hardware elements:

1. the Test Access Port (TAP),
2. the TAP Controller,
3. the Instruction Register (IR) and
4. a group of Test Data Registers (TDRs).

The first three elements and part of the group of TDRs are mandatory and *must* be supported by components which are compliant with the standard. The mandatory TDRs are the Boundary-Scan Register (BSR) and the Bypass register. The optional Test Data Registers comprise a Device Identification Register and one or more Design-Specific test data registers. Although the latter registers are optional, they must, if present in a component, obey the rules described in the IEEE Std 1149.1. The following sections detail each of these elements.

Figure 2-1 shows the standard Boundary-Scan architecture.
The operation of the Instruction and the Test Data registers is controlled by signals coming from the TAP Controller. The TAP controller is a sequential circuit which receives its controlling signals through the Test Access Port (TAP). The TAP only consists of IC contact pins. The standard TAP requires at least four IC pins (see also previous chapter) while a fifth pin may be provided for an optional active low reset signal TRST* (Test Reset), which can reset the test logic asynchronously to

the test clock signal (TCK). Usually, however, the reset operation is achieved through the TAP Controller.

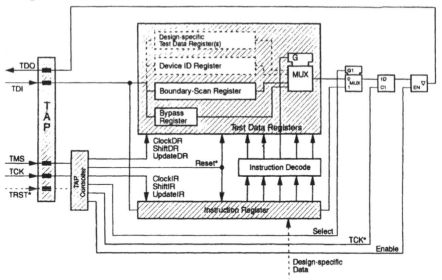

Fig. 2-1   The standard Boundary-Scan architecture
Note: The allowed optional registers and signals are shown in dotted lines.

The TAP Controller is driven by the signals Test Mode Select (TMS) and Test Clock (TCK), obtained from test execution hardware, supplied either externally from Automatic Test Equipment (ATE) or internally from system logic. More details are described in later sections.

Note in figure 2-1 that each of the named Registers can be selected into the serial signal path from TDI through TDO. The last multiplexer (MUX), as seen from TDI, selects either the Instruction Register or a Data Register. This multiplexer is directly controlled by the TAP controller. The preceding MUX selects which of the available data registers is to be switched into the TDI-TDO path, and is controlled by the decoded instruction.

TEST ACCESS PORT

The TAP provides access to the many test support functions built into an IC. It consists of four input connections, one of which is optional, and one output connection. The optional input connection is the TRST* pin.

The IEEE Std 1149.1 requires that the TAP connections are not used for any purpose other than testing.

The test signals connected to the Test Access Port have the following functions.

• *The Test Clock Input (TCK)*
  The TCK signal allows the Boundary-Scan part of the ICs to operate synchronously and independently of the built-in system clocks. A PCB may include various ICs that contain a component-specific clock. To remain independent of such clock frequencies, the TCK must *not* interfere with any system clock. However, for some optional tests a synchronization of the two clocks may be required. TCK permits clocking test instructions and data into or out of the register cells. Shifting in data from the TDI input pin must occur on the rising edge (positive slope) of the TCK pulse and shifting out data towards TDO occurs at the falling edge (negative slope) of TCK. Loading of data at the system input pins occurs at the rising edge of TCK.

• *The Test Mode Select Input (TMS)*
  The logic signals (0s and 1s) received at the TMS input are interpreted by the TAP Controller to control the test operations. The TMS signals are sampled at the rising edge (positive slope) of the TCK pulses. The signals are decoded in the TAP Controller to generate the required control signals inside the chip. When TMS is not driven it must be held at a logic 1. This may be obtained by a pull-up resistor on the TMS input pin in the TAP.

• *The Test Data Input (TDI)*
  Serial input data applied to this port is fed either into the instruction register or into a test data register, depending on the states of the TAP Controller. The input data is shifted in at the rising edge (positive slope) of the TCK pulses. When TDI is not driven, a logic 1 must be present. This may be obtained by a pull-up resistor on the TDI input pin in the TAP.

• *The Test Data Output (TDO)*
  Depending on the state of the TAP controller, the contents of either the instruction register or a data register are serially shifted out towards the TDO. The data out of the TDO is clocked at the falling edge (negative slope) of the TCK pulses. When no data is shifted through the cells, the TDO driver is set to an *inactive* state, for example to high impedance.

• *The Test Reset Input (TRST*)*
  The TAP's test logic is asynchronously forced into its reset mode when a logic 0 is applied to the TRST* pin. Therefore, a non connected TRST* pin must react as if a logic 1 was applied. TRST* is an *optional* signal, because in any case the test logic must be designed such that it can be reset under control of the TMS and TCK signals. A TRST* signal can be used to force the TAP

Controller logic into its reset state *independently* of the TCK and TMS signals at power-up time, as is required by the standard. The optional TRST* pin requires one more IC input pin for test purposes.

Figure 2-2 shows a board level configuration on which the TRST* is applied. At the same time this figure shows, as an example, an application where two TMS signals are used to test the upper and lower serial path independently (compare with figure 1-20). The latter is allowed because, as will be discussed later on in this chapter, when no data is shifted through the cells, the TDO driver is set to an inactive (high impedance) state.

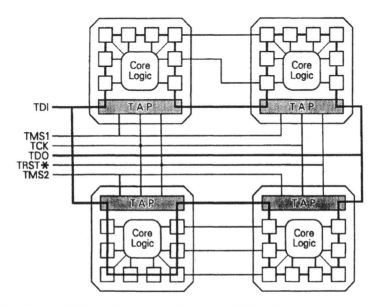

Fig. 2-2   BST configuration with two parallel paths

But care should be taken here, because the TMS signals must be controlled such that both the serial paths do not attempt to shift out signals simultaneously. A better practice in such cases is to use one serial net (figure 1-20) and sequentially set the ICs of one part of the serial path in the Bypass mode while testing the other part. The Bypass option is already mentioned in chapter 1 and is further discussed later on in this chapter.

TAP CONTROLLER

The basic function of the TAP Controller is to generate clock and control signals required for the correct sequence of operations of the circuitry connected to its output; that is, the Instruction Register and the Test Data Registers. Its main functions are:

- provide signals to allow loading the instructions into the Instruction Register,

- provide signals to shift test data into (TDI) and test result data out of (TDO) the shift registers,

- perform test actions such as capture, shift and update test data.

The operation of the TAP Controller is explained by means of its state diagram as defined by the IEEE Std 1149.1, see figure 2-3. All state transitions (as indicated by the arrows) within the TAP Controller occur on the rising edge of the TCK pulse, whereas actions in the connected test logic (registers etc.) occur at either the rising or the falling edge of the TCK. The values with the arrows are the TMS values at the rising edge (positive slope) of TCK.

Note that in figure 2-3 some states are shaded and others not. The non-shaded states are 'auxiliary' states, i.e. they do not initiate a system action but are included to provide 'process control'. For example if the test engineer wants to leave the *Capture-DR* state for an *Update-DR* action *without* a shift operation, then the path through *Exit1-DR* can be used. The *Pause-DR* state is included to provide a wait time during shifting to allow for example a test pattern generator to fetch/prepare test data. Shifting can be resumed by leaving the *Exit2-DR* state with TMS held at 0. These control sequences may cost one or two extra TCK cycles in the test procedure, but the alternative to provide these control functions at the cost of one or more extra test pins (in addition to TMS and TCK) at the IC package, which is not a reasonable option.

After switching on the system, it is essential to ensure that shortly after power-up the TAP Controller is in its *Test-Logic-Reset* state, allowing the system busses and wired junctions to be controlled by the system itself. As will be shown, the TAP Controller can be set synchronously into its *Test-Logic-Reset* state following five rising edges of the TCK, provided TMS is held at logic 1. However, the worst-case time required to reach this state may exceed the one at which damage could occur. It can also not be guaranteed that the clock will be running at the time at which power is applied.

These problems can be solved in various ways, such as:

- inclusion of a power-up reset within the ICs,

- inclusion of an asynchronous reset (TRST*) facility within the IC, by which the TAP Controller is reset asynchronously with respect to the TCK,

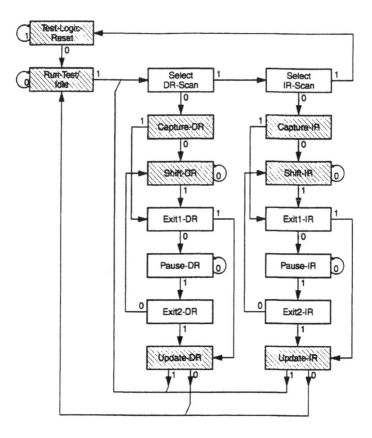

Fig. 2-3   State diagram of the TAP Controller

•   asymmetric design of the latches or registers used to construct the TAP
    Controller.

Where a power-up reset facility is provided within the component, it can be used
to initialize both the system and the test logic circuitry.

Now, suppose that after power-up using one of the above options the TAP
Controller has reached its *Test-Logic-Reset* state. As long as the TMS value is held
at 1, the controller remains in this state (see diagram). In order to leave this state,
the level of the TMS is set to 0. Then, at the next rising edge (positive slope) of the
TCK pulse, the TAP Controller leaves the *Test-Logic-Reset* state and enters the *Run-
Test/Idle* state. The *Run-Test* part of this name relates to the start of a built-in self-

test and the part *Idle* applies to all other cases. As long as TMS is held at 0, the controller remains in the *Run-Test/Idle* state (see figure 2-3).

The first step in the test procedure is to load an instruction, so the '-IR' branch (right column) of the state diagram must be entered. As a result the Instruction Register (IR) will be connected between the TDI and TDO pins. In the '-DR' branch of the state diagram (middle column) the test Data Registers (DR) operate. To select the '-IR' branch from the *Run-Test/Idle* state, two TCK pulses (rising edges) are needed with TMS kept at 1. With this action the *Select DR-Scan* state is passed. At board level all IRs are serially connected, once the TAP Controller is in the '-IR' branch. Once in the Select *IR-Scan* state, the *Capture-IR* and the *Shift-IR* states can be entered by making TMS a 0 for the next two TCK pulses. Shift actions are continued as long as the TMS line is kept at 0 in the *Shift-IR* state of the controller for as many clock pulses as required. When TMS is next set to 1, the shift process is ended and the next state *Exit-IR* is entered. When TMS is again set to level 1 the *Exit-IR* state in the controller is left and the *Update-IR* is entered. The instruction data are now clocked into the parallel output stages of the IRs and the new instructions become valid in the ICs. The 'Exit' states in the TAP Controller are temporary states, because whatever the TMS value is, this state is left at the next rising edge of the TCK pulse. In between the *Pause-IR* is present in the '-IR' branch of the state diagram. When in this state the TMS signal is kept at 0, the TAP Controller remains in this state. This is needed with multiple chains of different lengths. If after a pause the shift process must be resumed, the *Pause-IR* state is left with TMS at 1. Then *Exit2-IR* is entered which should be left then with TMS at level 0 to bring the controller back to its *Shift-IR* state.

After the *Update-IR* state the TAP Controller enters either the *Run-Test/Idle* state (TMS at 0) or directly to the *Select DR-Scan* state (TMS at 1). When the controller is in the '-DR' branch, the selected test data registers are serialized and similar actions as described here for the '-IR' can be initiated for the test data registers.

In this way the TAP Controller supports all test procedures that may be required by the test engineer. Figure 2-4 shows a timing diagram of the sequences described above, except that the *Pause-IR* is left via the *Exit2-IR* state.

Note in the state diagram (figure 2-3) that, in whichever state the TAP Controller is, a maximum of five TCK pulses will always reset the controller in its *Test-Logic-Reset* state, provided TMS is held at a logical 1.

The *non-shaded* states in figure 2-3 (*Select-DR-Scan*, *Select-IR-Scan*, *Exit1-DR*, *Exit1-IR*, *Pause-DR*, *Pause-IR*, *Exit2-DR* and *Exit2-IR*) are used to control the required test sequence. The *Pause* is provided to temporarily halt the shifting process. This state can be used, for example, to allow synchronization between TCK and system clock signals, when needed for certain tests. These 'Pause' states are maintained as long as the level of TMS is kept low (logical 0). Further to that, the

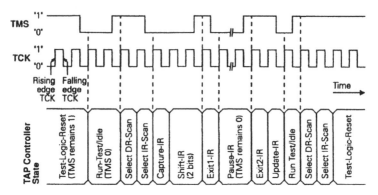

Fig. 2-4   Example timing diagram of TAP Controller

state *Exit1* is present in order to provide a three way branch starting from *Capture* to *Shift* or (via the *Exit1*) to either *Pause* or *Update* just by using one control signal (TMS). In the same way *Exit2* is provided to allow a three way branch starting from *Pause* to either stay in the *Pause* state or leave *Pause* (via the *Exit2*) to either *Update* or *Shift*. So with this construction a three way branch is controlled with one signal (TMS), for which two control signals were required otherwise.

It should be noted that when an *Exit* state is left with the TMS at a logical 1, an *Update* action takes place, meaning that the scanning process has terminated.

In the following subsection the operation of all possible TAP Controller states are described, following the formal specifications.

State Descriptions

All controller states shown shaded in the state diagram (figure 2-3) control the *changes* of the connected test logic. The resulting signal flows through the Boundary-Scan Cell and the Instruction Register are described below, using figures 1-21 and 1-22 respectively.

• *Test-Logic-Reset*
   In this controller state (see figure 2-5) all *test* logic is disabled, i.e. all core system logic operates normally. Whatever the state of the controller is, it will enter the *Test-Logic-Reset* state when the TMS signal is held high (logic 1) for at least five rising edges of the TCK pulse. The exact required number of TCK pulses depends on the current state of the TAP Controller. The controller remains in this state while TMS is held high. If the optional signal TRST* is

present in the TAP, it can be used to force the controller to the *Test-Logic-Reset* state at once (asynchronous reset), at any moment during circuit operation.

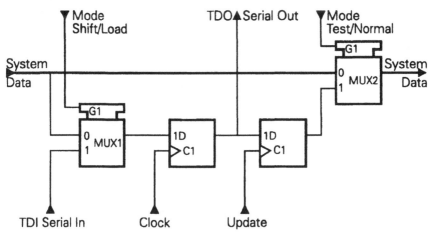

Fig. 2-5   Signal flow (bold lines) during Test-Logic-Reset

• *Run-Test/Idle*

This is a controller state between the various scan operations, see previous section. Once the controller is in this state, it will stay there as long as the TMS signal is held low (logical 0). The operation of the connected test logic depends on the instruction contained in the instruction register. If the instruction causes a built-in self-test function to execute (e.g. the RUNBIST instruction) the TAP Controller has to wait for its completion. This can be done by holding the TAP Controller in the Run-Test/Idle state. If an instruction does not cause built-in self test functions to execute, such as select a data register to scan, then all selected test data registers will stay *Idle* while the TAP Controller is in the Run-Test/Idle state. The current instruction does not change while the controller is in this state.

• *Capture-DR*

In this controller state (see figure 2-6), data is parallel-loaded from the parallel inputs into the selected test data register. The register retains its previous state if it does not have a parallel input or if capturing is not required for the selected test. The action takes place at the rising edge (positive slope) of the TCK pulses.

• *Shift-DR*

In this controller state (see figure 2-7), the previously captured data is shifted out towards the TDO, one shift-register stage on each rising edge (positive slope) of the TCK pulse. If a selected test data register is not placed in the serial path, it retains its previous state.

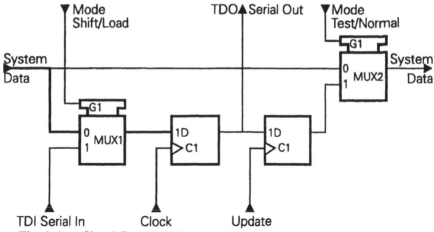

Fig. 2-6   Signal flow (bold lines) in BS cell during Capture-DR

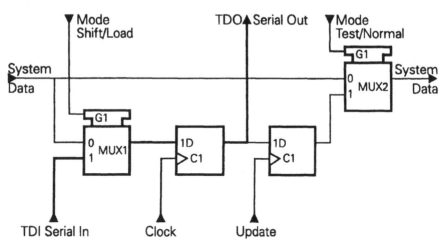

Fig. 2-7   Signal flow (bold lines) in BS cell during Shift-DR

Update-DR
Once the controller is in this state (see figure 2-8), the shifting process has been completed.Test data registers may be provided with a latched parallel output. This prevents the parallel output from changing while data is shifted into the associated shift-register path. When these test data registers are selected by an instruction, the new data is latched into their parallel outputs in this state at the

falling edge (negative slope) of the TCK pulses. The shift-register stage in a selected test data register retains its previous state.

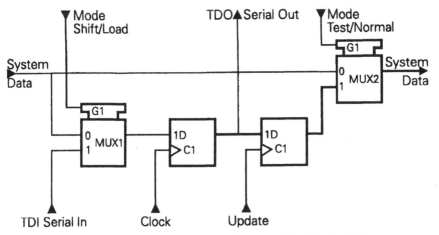

Fig. 2-8   Signal flow (bold lines) in BS cell during Update-DR

*   *Capture-IR*

In this controller state (see figure 2-9), the previously shifted-in instruction data is parallel-loaded into the shift-register stage of the instruction register. In addition, design-specific data may be loaded into a shift-register stage which may not to be set to a fixed value. In this way the IR-path between TDI and TDO on the board (see figure 2-12) can be checked as to whether or not the instructions can be shifted in correctly or not. The action takes place at the rising edge (positive slope) of the TCK pulses. Test data registers which are selected by the current instruction retain their previous state.

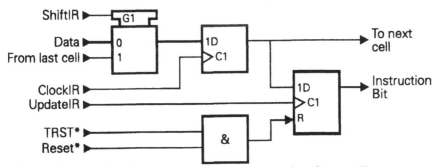

Fig. 2-9   Signal flow (bold lines) in IR cell during Capture-IR

*   *Shift-IR*

In this controller state (see figure 2-10), the previously captured data is shifted out towards the TDO, one shift-register stage on each rising edge (positive slope) of the TCK pulse. Selected test data registers retain their previous state.

The current instruction does not change in this controller state.

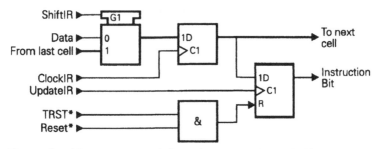

Fig. 2-10   Signal flow (bold lines) in IR cell during Shift-IR

* *Update-IR*

  The shifted-in instruction data is loaded from the shift-register stage into the
  parallel instruction register (see figure 2-11). The new instruction becomes valid
  when the TAP Controller is in this state. All the test data shift-register stages
  which are selected by the current instruction retain their previous state. The
  action takes place at the falling edge (negative slope) of the TCK pulses.

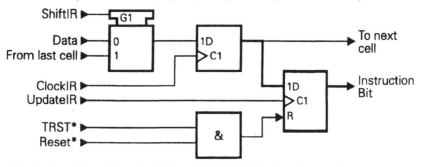

Fig. 2-11   Signal flow (bold lines) in IR cell during Update-IR

THE INSTRUCTION REGISTER

The Instruction Register allows test instructions to be shifted into every IC along
the PCB-level path. Therefore, in the Shift-IR controller state, all IC instruction
registers are connected in series at board level between TDI and TDO. A test
instruction defines the test data register to be addressed and the test to be
performed. This may vary per IC on the board, as is depicted in figure 2-12.

The IEEE Std 1149.1 defines a number of mandatory and optional instructions.
Further design-specific instructions may be added by the designer of the IC.

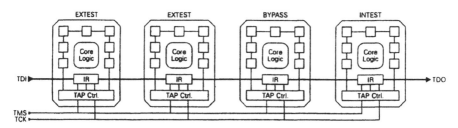

Fig. 2-12   ICs set for various tests

The design of the Instruction Register is a serial-in parallel-out register (see figure 1-22). Each Instruction Register cell has a shift-register flip-flop and a parallel output latch. The shift-register flip-flop holds the instruction bit moving through the Instruction Register. The parallel output latch holds the current instruction bit, latched into it from the shift-register during the *Update-IR* state. The latched instruction can only be terminated/changed when the TAP Controller is in its *Update-IR* or *Test-Logic-Reset* state. No inversion of data between the serial input and the serial output of the Instruction Register is allowed to take place.

The Instruction Register must contain at least two shift-register-based cells which can hold instruction data. These two mandatory cells are located nearest to the serial outputs, i.e. they are the least significant bits. Figure 2-13 shows the set-up of an Instruction Register.

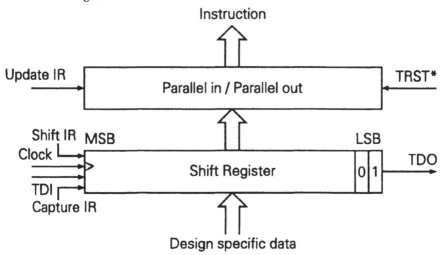

Fig. 2-13   Set-up of an Instruction Register

The values of the bits are 0 and 1 (the 1 is the least significant bit). These fixed values are used in locating faults in the serial path through the ICs on the PCB. The fixed binary "01" pattern must be loaded in the *Capture-IR* state of the TAP Controller.

Additional parallel inputs are allowed to permit capture of design-specific information. In cases where no design-specific instructions are defined, it is recommended that the design-specific cells are designed such that the parallel inputs load a fixed logic value (0 or 1) in the *Capture-IR* state of the TAP Controller.

Table 2-1 shows the operation of the Instruction Register (IR) in each of the states of the TAP Controller.

Table 2-1   IR Operation in Each TAP Controller State

Controller state	Shift-Register Stage	Parallel output
Test-Logic-Reset	Undefined	Set to give the IDCODE (or BYPASS) instruction
Capture-IR	Load 01 into LSBs and design-specific data or fixed values into MSBs	Retain last state
Shift-IR	Shift towards serial output	Retain last state
Exit1-IR	Retain last state	Retain last state
Exit2-IR	Retain last state	Retain last state
Pause-IR	Retain last state	Retain last state
Update-IR	Retain last state	Load from shift-register stage into decoder
All other states	Undefined	Retain last state

Referring back to the description of the TAP Controller state diagram along with figure 2-3, it can be noted from the table that the states *Exit1-IR, Exit2-IR* and *Pause-IR* do not influence the operation of the test logic.

TEST DATA REGISTERS

The Boundary-Scan Architecture must contain a minimum of two test data registers: the Bypass Register and the Boundary-Scan Register. A third register, the Device Identification Register, is optional. One or more additional design-specific test data registers are allowed too. The design-specific registers may, but need not, be

publicly addressable. Additional registers are accessed by adding instruction codes
to the architecture's test-instruction set.

Each test data register included in the architecture has a fixed length as designed by
the IC designer and can be accessed through one or more instructions that can be
shifted into the Instruction Register. Figure 2-14 shows a detail of the standard
Boundary-Scan Architecture (compare with figure 2-1).

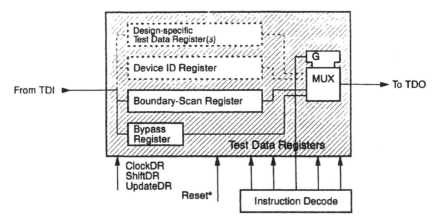

Fig. 2-14   Overview of the Test Data Registers
Note: The allowed options are shown in dotted lines.

The depicted Test Data Registers are described in more detail in the following
sections.

It should be mentioned here that the length of a test data register must be fixed
(defined by the IC designer), independent of the instruction by which it is accessed.
A named test data register may share registers from other logic circuitries, but each
part and each addressable combination of parts must have a unique name and the
resulting design must be compatible with the IEEE Std 1149.1.

Figure 2-15 shows an example.

Table 2-2 shows the operation of the Test Data Registers in each of the states of the
TAP Controller.

Referring back to the description of the TAP Controller state diagram along with
figure 2-3, it can be noted from the table that the states *Exit1-IR, Exit2-IR* and
Pause-IR do not influence the operation of the test logic.

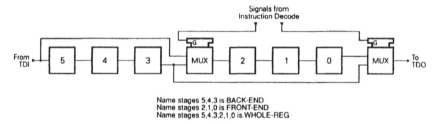

Name stages 5,4,3 is BACK-END
Name stages 2,1,0 is FRONT-END
Name stages 5,4,3,2,1,0 is WHOLE-REG

Fig. 2-15   Example of a combined test data register

Table 2-2 Operation of the Test Data Register Enabled for Shifting

Controller state	Action
Capture-DR	Load data at parallel input into shift-register stage. Parallel output register or latch retains last state.
Shift-DR	Shift data towards serial output. Parallel output register or latch, retains state.
Exit1-DR	Retain last state.
Exit2-DR	Retain last state.
Pause-DR	Retain last state.
Update-DR	Load parallel output register or latch from shift-register stage. Shift-register stage retains state.
All other controller states	Registers which have a parallel output maintain their last state of the output; otherwise undefined.

The Bypass Register

The mandatory Bypass Register is a single stage shift-register. When selected, the shift-register is set to a logic zero on the rising edge (positive slope) of the TCK with the TAP Controller in its *Capture-DR* state. It then provides a minimum length serial path for the test data shifting from the component's (IC) TDI to its TDO.

The benefit of the Bypass Register is a shortened scan path through the Boundary-Scan architecture when scan access of the test data registers is not required, or it can be used to leave the IC in functional mode during testing. Chapter 1 also describes some applications of the Bypass register.

Suppose a printed circuit board has 50 ICs and the Boundary-Scan registers of all ICs are connected into a single serial chain. If the average component register length is 100 stages, then the total serial path length contains 5000 stages. If only one IC in this chain is to be tested (with the INTEST instruction) then the remaining 49 ICs can be bypassed. For this test of one IC, only 100 stages plus 49 bypass stages are required. Now, instead of all 5000, only 149 (about 3%) of the stages are involved. This reduces testing time considerably in two ways:

1. The test cycle is shortened, in this example, by 97%.

2. Only the 100 bits for that one IC under test are of importance. Hence, the rest of the bits need not be stored in the automatic test equipment, which shortens diagnosis times.

Figure 2-16 shows an example of an implementation of a Bypass Register or cell.

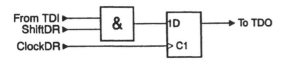

Note that the register does not have a parallel data output latch.

Fig. 2-16   Implementation of a Bypass Register

When the Bypass Register is selected, the shift-register stage is set to a logical zero on the rising edge (positive slope) of the TCK pulse when the TAP Controller is in its *Capture-DR* state. This allows to test the presence or absence of the optional device identification register. When the device ID is present, then a constant logic 1 must be loaded in its LSB. When the *IDCODE* instruction (see the next section) is loaded into the Instruction Register, then a subsequent data register scan cycle causes the first data bit of each component to be shifted out. So when this bit is a 1, then the optional device ID register is present, otherwise the instruction selects the Bypass register, out of which a logic 0 is shifted.

The Boundary-Scan Register

The mandatory Boundary-Scan Register (BSR) consists of a series of Boundary-Scan Cells (BSCs) arranged to form a scan path around the core logic of the IC (see figure 1-19). It provides the following features:

• It allows testing of circuitry external to the component (IC). A typical example is the interconnect test. Also logic circuitry embedded between Boundary-Scan testable devices (RAM, ROM, glue logic) can be tested, using the *EXTEST* instruction.

- It allows testing of the core logic (INTEST), while also providing defined conditions at the periphery of the core logic.

- It allows sampling and examination of the input and output signals without interfering with the operation of the core logic.

- And last but not least: it can stay idle, thereby showing virtually no load to the core logic when signals are flowing through the system.

Chapter 1 describes how the test data flow in a component (IC) is controlled by the Boundary-Scan Register for various instructions and figure 1-18 shows the interconnections between the Instruction Register, the Boundary-Scan cells and the TAP Controller.

A Boundary-Scan Cell

For ease of understanding, an example of a 'universal' Boundary-Scan Cell (BSC) is considered. The example is not prescribed by the IEEE Std 1149.1 standard nor is it considered here as a preferred configuration. But it is a configuration that could be used at both the input and the output pins of an IC. However, not all elements of the cell are always needed in a particular application and the building blocks composing the cell can also be different.

A Boundary-Scan register cell must be dedicated to the test logic only, i.e. it shall never be used for system operation. In order to meet the requirements of all instructions, it must be possible in a BSC to move test data through the register without interfering with the normal system operation. This can be achieved by making the shift-register stages and their parallel output latches a dedicated part of the cell.

Figure 2.17 shows an implementation of a Boundary-Scan Cell which can be used at both input and output pins. This configuration is often referred to in publications. However, it is certainly not the best one on all occasions. A simpler cell can for example be used at the input pins, if one follows strictly the minimum requirements of the IEEE Std 1149.1, see figure 2-18. This cell may be preferred over the previous one in delay sensitive circuits, for example at clock inputs.

The minimum requirement follows from the mandatory instructions as prescribed in the IEEE Std 1149.1: the *EXTEST* and the *SAMPLE/PRELOAD* instructions. So for the input cell no more logic elements are necessary, because the test results from a PCB interconnect test with the *EXTEST* instruction can be captured and further shifted towards the TDO pin. Obviously, the test of core logic of an IC which is mounted on a PCB is an option in the standard, for which the *INTEST* instruction is prescribed if it is supported by the chip manufacturer.

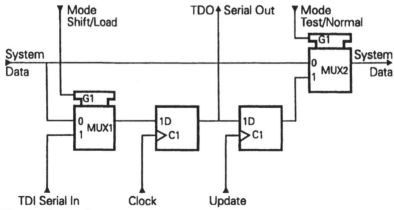

Fig. 2-17   An implementation of a Boundary-Scan Cell

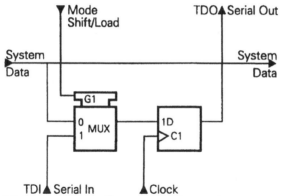

Fig. 2-18   Minimum BSC configuration for input pins

At the output pins, the cell depicted in figure 2-17 fulfils the minimum requirement.

System Pin Related Cells

Next to the mandatory overall specifications of the Boundary-Scan Register cells, additional specifications are valid for cells connected to the various types of *component or system pins*. Figure 2-19 shows schematically how the Boundary-Scan Cells surround the core logic and provide signals for the various output buffers. Control of the BSCs is provided by the TAP Controller. Some details concerning the various output cells are given below.

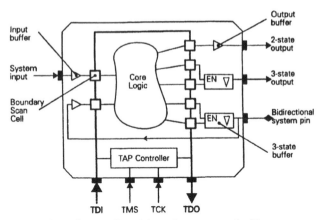

Fig. 2-19   Boundary-Scan cells with various output buffers

Notice that the discussed cells may also transmit signals coming from the purely digital core logic into a mixed analog/digital circuit block.

Note: In the descriptions below various instruction names are mentioned. These instructions are explained in a subsequent section.

• *Cells at 2-State Output Pins*
At 2-state output pins the signals can be set only at a high or a low logical level at any given instant. Therefore, only one Boundary-Scan cell per pin is sufficient to observe or control the state of the pin (see figure 2-19). Problems may be encountered when Boundary-Scan cells drive signals into logic blocks external to the component. Such an external block may contain asynchronous logic that will be set into undesirable states when shifting patterns appear at its input. Also, Boundary-Scan output signals may be fed into a clock input of the external block, which may produce hazardous effects if the logic is not shielded from the shifting patterns.

Therefore, the Boundary-Scan cell should be designed such that these problems are guaranteed to be avoided. The example cell in figure 2-17 fulfils this rule. In particular the parallel output register or latch can affect the state of an output driver at a system pin as in figure 2-19. This ensures that, while the *EXTEST* instruction is selected, the data driven from a component to an external logic block changes only after completion of the shifting process, at the update state transition.

• *Cells at 3-State Output Pins*

At 3-state output pins the signals can be actively set at a high or a low logical level *and* the pin can be held at an inactive state. Therefore, at least two Boundary-Scan cells per pin are needed to observe or control the state of the pin (see figure 2-19).

The 3-state system output pin may drive a wired junction on PCB level, using the *EXTEST* instruction. Therefore it must be possible to drive the wired junction from each of the possible driving pins independently. To that purpose the 3-state and bidirectional output drivers can be used.

At PCB level the output pin may also drive a component of which the core logic is being tested using the *INTEST* or *RUNBIST* instruction. Two options are available to avoid contention in such cases:

1. The state of the system pin can be fully user defined like a 2-state output pin.

2. The system pin can be driven into its inactive high-impedance state. This is allowed since components must tolerate high-impedance conditions during normal system operation.

Figure 2-20 shows the details for such a configuration.

The internal connections between the Boundary-Scan Cells may be as follows: The control lines of both cells (*Shift-DR*, *Clock-DR*, *Update-DR* and *Mode*) are connected in parallel. The TDO signal of the *Output cell* is connected to the TDI of the *Control cell*. In this way the TDI/TDO serial shift-register path is maintained. But other internal connections between the two cells are also possible.

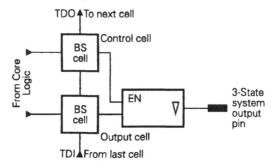

Fig. 2-20   Cell configuration at a 3-state output pin

Note: When two more or more 3-states pins are present (e.g. connected to a
         bus) it is allowable to control the respective enable stages with one
         Boundary-Scan cell. Figure 2-21 shows an example.

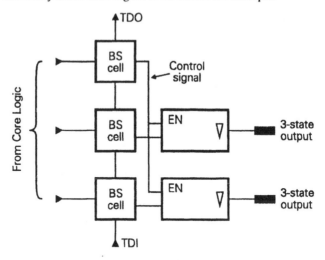

Fig. 2-21   One BS cell controls several 3-state outputs

• *Cells at Bidirectional Pins*
  Bidirectional pins may be either a 2-state (for example formed as a combination
  of an open-collector output and an input) or a 3-state pin (for example formed
  as a combination of a 3-state output and an input). As a matter of fact, the IEEE
  Std 1149.1 requirements are a merge of those for both the 2-state or 3-state
  output pins and the system input pins.

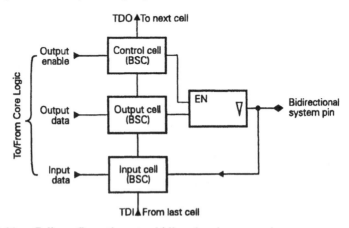

Fig. 2-22   Cell configuration at a bidirectional system pin

Figure 2-22 shows the details of a configuration at a 3-state output pin. The internal connections between the Boundary-Scan Cells for the bidirectional output pin are the same as in the 3-state output pin, except that the parallel output flip-flop of the control cell may have an extra input for the *Reset** signal.

Note: One output cell for several pins is allowed.

The Device ID Register

The Device Identification Register is an optional register that may be included in an IC. But *if* it is included, it must follow the requirements described in the IEEE Std 1149.1.

The Device ID Register provides binary information about the manufacturer's name, part number and version number of the component (IC). This is important, for example for the following applications.

*   In the factory, it allows verification that the correct IC has been mounted on the proper place on the PCB.

*   When a component (IC) has been replaced, the version number of the replacement can be checked, and if required the test program can be modified.

*   It may be desirable to blindly interrogate a PCB design by a controller unit in order to determine the type of each component on each board location without further functional knowledge of the design.

*   When a PCB is added to a configuration at system level, the system test program can be adjusted to the new PCB and the ICs mounted on it.

*   The correct programming of off-line programmed ICs can be checked.

The register must consist of a 32 bits shift-register, parallel-in and serial-out. The register does not need a parallel output latch. Like the bypass register, the normal operation of the IC can continue while the ID register is in use.

The structure of the data loaded into the Device ID Register in response to the *IDCODE* is shown in figure 2-23.

The least significant bit (LSB) is always loaded with a logical 1. This is done in conjunction with the Bypass Register, which is always set to a logical 0. If the optional Device ID Register is not present, the Bypass Register is chosen when an *IDCODE* instruction is executed. If subsequently the first bit shifted out of the component during a test data scan is a 0 (zero), then it can be deduced that the

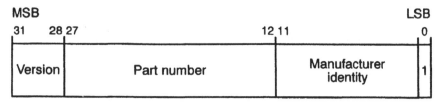

Fig. 2-23   Structure of a Device Identification Register

component has no Device ID register. The bits 1 through 11 contain an 11 bits manufacturer code, which is derived from a scheme controlled by the Joint Electron Device Engineering Council, abbreviated as JEDEC [3]. This scheme allows 2032 manufacturer identifications. The IEEE Std 1149.1 prescribes that the manufacturer code *00001111111* is not permitted for components claiming compatibility with the standard. This code can then be used as a known invalid code in test algorithms. The bits 12 through 27 allow 2^{16} ($\approx$ 65,000) part number codes. The remaining bits 28 through 31 permit 16 variants of a part to be distinguished, for example between chips showing a different behaviour on instructions or have different responses to data sent or received through the TAP.

The need for interrogation of the board component becomes evident when on a board components are used that can be user programmed for certain functions. Another application may be in systems which can be configured with boards for various functions, possibly supplied from different manufacturers.

Figure 2-24 shows an example configuration of a Device ID Register cell.

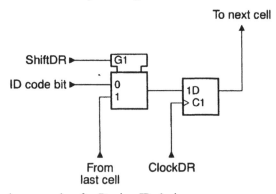

Fig. 2-24   An example of a Device ID design.

Design-Specific Registers

The manufacturer may add test data registers dedicated to his own design. For this purpose, design-specific instructions have been developed. In figure 2-1 at the beginning of this chapter, these options are shown in dotted lines.

One point should be made here. The manufacturer may decide whether he makes the instructions for the design-specific register available in his component catalogue or not. If not, he needs the design-specific registers only for his own in-house testing.

INSTRUCTIONS

Instructions are serially loaded from the TDI into the test logic during an instruction-register scan cycle, with the TAP Controller in its *Shift-IR* state. Subsequently the instructions are decoded to achieve two basic functions.

1. Select the set of test data registers that may operate while the instruction is current. Non-selected test data registers must be controlled so that they do not hamper the normal on-board operation of the related ICs.

2. Define the serial test data path that is used to shift data between TDI and TDO during data register scanning.

Instruction codes that are not used to provide control of test logic must be equivalent to the *BYPASS* instruction (described later). This makes sure that a test data register is always connected between TDI and TDO for every possible instruction code. So every instruction code will produce a defined response.
The IEEE Std 1149.1 specifications of the instructions cover:

• the instructions, whether they are mandatory or optional,

• test data registers which are to be connected in the serial TDI/TDO path,

• binary code patterns (if applicable), and

• the data flow through Boundary-Scan Register cells and/or core logic signals.

Public and Private Instructions

Public instructions are documented and supplied by the manufacturer along with the component. These instructions provide the component user with access to the test features, so that the user can apply the instructions in test tasks on component, PCB or system level.

In addition to the public instructions, private instructions exist, which are available solely for the use of the component manufacturer. The operation of private instructions need not be documented. If private instructions are utilized in a component, it remains the responsibility of the vendor to clearly identify those instruction codes that, if selected, will damage the component or cause hazardous operation of the component. The following subsections describe the public Boundary-Scan instructions.

The BYPASS Instruction

The *BYPASS* instruction is the only instruction defined by the IEEE Std 1149.1 that can be applied when the Bypass Register is to be selected in the TDI-TDO path. The binary code for the *BYPASS* instruction must be {111...1} or said as 'all 1s'. The use of this register and the purpose of the code is already described in previous sections, including chapter 1. The instructions can be entered by holding the TDI at the logical high level.

Note that when the TDI is not terminated (e.g. caused by an open on the board level), it behaves as if the input was high (logical 1). This causes the Bypass Register to be selected following an instruction-scan cycle, so avoiding any unwanted interference with the normal IC operation.

If the optional Device ID Register is not present, then the *BYPASS* instruction is forced into the latches at the parallel outputs of the Instruction Register when the TAP Controller is in its *Test-Logic-Reset* state. This ensures a complete serial TDI/TDO path in either case.

The SAMPLE/PRELOAD Instruction

This is a mandatory instruction used to scan the Boundary-Scan register without interfering with the normal operation of the system core logic. Received system data at the input pins is supplied to the core logic without modification and system output data is driven to the output pins without interference. Meanwhile samples of these signals can be taken and the sampled data can be shifted through the Boundary-Scan registers.

So the instruction supports two functions:

1. It allows a *SAMPLE* ('snapshot') of the normal operation of a component (IC) to be taken for examination.

2. Prior to the selection of another test operation, a *PRELOAD* can take place of data values into the latched parallel outputs of the Boundary-Scan cells.

The *SAMPLE* mode is a useful tool for debugging of prototypes in the development phase of a board design. In the sampling mode data are captured with the TAP Controller in its *Capture-DR* state at the rising edges of the TCK pulses. Subsequently these data can be shifted out in the *Shift-DR* state of the controller. The *PRELOAD* is selected when the user prepares an *EXTEST* by shifting in beforehand the data which must be driven out from the chip's output pins into the PCB net using the TAP controller in its *Update-DR* state.

An application of the PRELOAD function is the following. Directly after power-up the data in the Boundary-Scan registers at the output cells are not known. If, as a first test step, an EXTEST instruction is loaded into the Instruction Register, then the execution of the test may result in hazardous signals being output to the connected inputs on the net. Therefore, the Boundary-Scan registers at the output (driving) pins should be loaded beforehand with known data using the PRELOAD instruction.

Figures 2-25 and 2-26 show the input and output cells when the *SAMPLE* respectively the *PRELOAD* instructions is selected. The data flows in the Boundary-Scan cells are shown in bold lines.

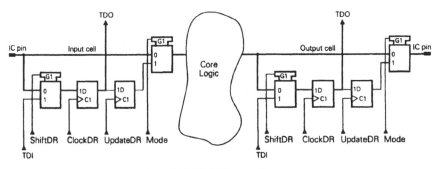

Fig. 2-25   Data flow during SAMPLE instruction

The EXTEST Instruction

Since the board-level interconnect testing is one of the primary reasons for introducing Boundary-Scan testing, the *EXTEST* is a mandatory instruction within the IEEE Std 1149.1. This is most important and specifically provided to allow board-level testing of *opens, stuck-at* or *bridging errors* etc. This instruction also allows testing of clusters of components which do not themselves incorporate

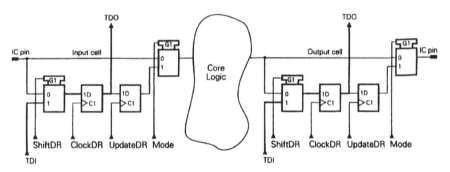

Fig. 2-26   Data flow during PRELOAD instruction

Boundary-Scan Test features but are surrounded by Boundary-Scan components. Cluster testing is discussed further in chapter 5.

When the *EXTEST* instruction is selected, the core logic is 'isolated' from the input and output pins. The test data for the execution of the *EXTEST* instruction is loaded beforehand into the Boundary-Scan Register stages using the *SAMPLE/PRELOAD* instruction. The loading of test stimuli is concluded by bringing the TAP controller to the *Update-DR* state. On the falling edge of the TCK pulses in this state, the test stimuli are transferred to the parallel output stage. The *EXTEST* instruction has caused the Mode input of this output stage to 1 (see figure 2-27), so that the test stimuli are directly transferred to the PCB net. At the receiving end(s) of the net, the cells at the input pins capture the test result with the controller in its *Capture-DR* state. The next test step in this procedure shifts out the test results from the input pin cells towards TDO for examination. Please refer also to the description of the *EXTEST* in chapter 1.

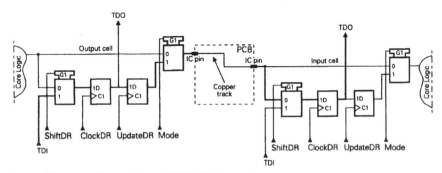

Fig. 2-27   Data flow during EXTEST instruction

Finally, care must be taken if more output pins are connected to the same net, for example, a bus. It is recommend that during the time of the execution of the *EXTEST* instruction only one system pin is driving a net at a time, while the other connected output pins are kept at HIGHZ. This avoids Boundary-Scan Cells at the output pins being back-driven (overdriven) with an unknown signal value, which may cause damage to the component.

The INTEST Instruction

When the *INTEST* is selected, the Boundary-Scan test substitutes testing by means of the bed-of-nails fixture and pin probing of automatic test equipment (ATE) for component (IC) testing.

This is an optional instruction which, if provided, must comply with the IEEE Std 1149.1. The INTEST instruction allows static (slow-speed) testing of the core logic after the component is mounted on the PCB. Following this instruction, test signals are shifted in one at a time and applied to the core logic. As explained in chapter 1, the operation resembles the *EXTEST*; only the role of the IC's output and input pins is reversed.

Figure 2-28 shows the data paths through the Boundary-Scan Cells applied at the input for the *INTEST* and shifting out the test results from the output side of the core logic.

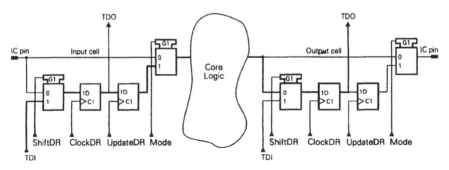

Fig. 2-28   Data flow during INTEST instruction

As stated above, the test signals are applied one at a time at the rate of TCK. This requires that the core logic can operate at this rate. For devices containing dynamic logic, such as DRAM memories, refreshment of the data cells may require a much higher frequency than can be obtained with this test method. Unless core logic provisions have been taken internally (e.g. synchronization of the system clock with TCK), this test method could be unusable for these devices. Therefore the *RUNBIST* may be best suited (see next subsection).

Between the input of the signal and the capture of the response, the core logic may receive clock signals. These clock signals may be obtained in various ways while the *INTEST* instruction is executed. Some examples are given in the description of the IEEE Std 1149.1.

The RUNBIST Instruction

The Run Built-In Self-Test is an optional instruction which, if provided, must comply with the IEEE Std 1149.1. The *RUNBIST* instruction causes the execution of a self-contained self-test of the component (IC) without the need to apply complex data patterns. With the *RUNBIST* instruction the correct operation of a component (IC) can be tested dynamically, without the need to load complex data and without the need for single-step operation (as required for the *INTEST* instruction). The IEEE Std 1149.1 requires that this self-test can only be executed with the TAP Controller in its *Run-Test/Idle* state. Only then the self-test can run to completion, provided that the controller remains long enough in the *Run-Test/Idle* state, measured for example in clock pulses applied to the core logic. The result of the self-test is captured from the selected test data register with the controller in its *Capture-DR* state, after which it can be shifted out for examination.

The timing constraints have been added to ensure that the built-in self-tests of *all* components involved are completed in one test run. If on a PCB components are mounted with different specifications of self-test duration times, then the IEEE Std 1149.1 ensures that enough time has passed for the completion of the built-in self-tests of all components.

When the self-test is running, the Boundary-Scan cells are used to hold the component's outputs to a fixed value. In this way the signals generated in the core logic during the self-test can not enter the PCB nets, so that unwanted values at the receiving ends of the nets can not damage the connected logic.

Finally it should be mentioned that, when the *RUNBIST* instruction is applied, the test results for all versions of a component must be the same.

The CLAMP Instruction

The optional *CLAMP* instruction [4] is used to control the output signal of a component to a constant level by means of a Boundary-Scan cell. In such cases the Bypass Register is connected in the TDI-TDO path on the PCB. The instruction is used for instance with cluster testing, where it can be necessary to apply 'static' guarding values to those pins of a logic circuitry which are *not* involved in a test. The required signal values are loaded together with all test stimuli, both at the start of the test and each time a new test pattern is loaded. As such the *CLAMP* instruction does increase the test pattern, thus slightly reducing the overall test rate.

The IDCODE Instruction

If a Device Identification Register is included in a component, the *IDCODE* is forced into the Instruction Register's parallel output latches while the TAP Controller is in its *Test-Logic-Reset* state. This means of access to the Device Identification Register permits blind interrogation of components assembled onto a PCB, making it possible to determine what components are mounted on a board. The possible applications of the Device ID register are described above in the relevant subsection.

The USERCODE Instruction

The *USERCODE* must be provided by the manufacturer if the Device Identification Register is included in a component *and* the component is user-programmable. This instruction is only required if the programming can not be determined through the use of the test logic. When selected, this instruction loads the user-programmable identification code into the Device Identification Register, at a rising edge of TCK and the TAP Controller in its *Capture-DR* state.

The HIGHZ Instruction

The optional *HIGHZ* instruction [4] forces *all* output of a component to an inactive drive state, for example to a high impedance state. The application of the *HIGHZ* instruction is found in situations where besides Boundary-Scan Test, a conventional in-circuit test is still required, e.g. for testing non-BST compatible components. The in-circuit tester may drive signals back to the component's output pins where hazards may occur if its output impedance is not high.

DOCUMENTATION REQUIREMENTS

Manufacturers can claim that their components comply with the IEEE Std 1149.1 for Boundary-Scan testing. Such components must comply with all relevant specifications described in the standard. The level of conformance to the standard may vary, however, depending on the range of test operations that is supported. The claim must clearly identify the subset of the public instructions that is supported (see next table).

Table 2-3   Public Instructions

Instruction	Status
BYPASS	Mandatory
SAMPLE/PRELOAD	Mandatory
EXTEST	Mandatory
INTEST	Optional
RUNBIST	Optional
IDCODE	Optional
USERCODE	Optional

The minimum requirement for conformance with the IEEE Std 1149.1 is set to ensure that the IC user can perform two basic test tasks:

1. examine the operation of a prototype system (e.g. IC), and

2. test assembled products for assembly-induced defects during manufacturing.

Further it is strongly recommended that the components support either the *INTEST* or the *RUNBIST* instruction or both, in order to enable efficient and comprehensive verification of internal component operation at the board and system level.

ASIC vendors may claim conformance with the IEEE Std 1149.1 if they can prove that an interconnection of cells, if made, will produce a component that meets the requirements of the standard.

With the sole exception of the Device Identification code, each second source component must react on all public instructions in exactly the same way as the components from the prime source supplier. This ensures exchangeability of components on a PCB, which is beneficial with respect to prepared test programs in the manufacturing phase and in field service.

The operation of the component's test logic must be fully documented if conformance is to be claimed for that component. The information, specified in the standard, must be supplied by the component supplier. The component purchaser needs this information for use in development tests, production, field service and other activities.

Chapter 3

HARDWARE TEST INNOVATIONS

This chapter describes provisions that can be made in order to support Boundary-Scan testing for printed circuit boards (PCBs) and a system. A number of on-chip provisions are also illustrated.

PROVISIONS AT BOARD LEVEL

This section describes some innovations that are primarily introduced to improve the quality of loaded or assembled board testing in terms of speed and fault coverage. The descriptions refer to available or proposed products and test methods from various companies. It is not the intention to be complete in this section but the examples are meant to give a reference for those who want to start applying Boundary-Scan testing of boards in their company.

Concurrent Sampling

Heavily loaded printed circuit boards (PCBs) may contain so many components that it becomes very difficult to ensure that all devices function properly under all envisaged circumstances. Such boards may contain microprocessors with co-processors and connecting buses, or memory units with accompanying DMA/memory controller. Very large volumes of test code have to be developed by various design groups that participate in the design of the PCB. The final code should then preferably be suited for board testing (assembly and functional tests), system testing and for field service environments. Concurrent sampling for boards using self-test code is introduced by IBM [5]. Test code is enhanced by concurrently sampling signals at chip boundaries, compressing this data and verifying its signature in-line in the code.

Before going into details and showing the advantages of concurrent sampling, some relevant properties of the traditionally used assembly and functional test methods are summarized.

Assembly testing should be performed before functional testing, because the majority of board defects (opens, shorts or wrong components) are assembly related. Boundary-Scan test technology may be used very efficiently in conjunction with analogue structural testing such as MDA (Manufacture Defect Analysis). It allows individual chip tests and testing of clusters of ICs by isolating them from the rest of the PCB circuitry. This assembly test usually covers 95% of the total board defects. The remaining defects are caused by timing/skew errors at board level or by defective ICs that are not detected by INTEST, RUNBIST or cluster test.

Functional tests are performed by applying and observing signals to the PCB (edge) connector pins. The test signals should be as close as possible to the expected signals during normal system operation, from static to full speed. With boards containing microprocessors, the code of the functional tests may include code of functional self tests (FSTs) executed by the microprocessor(s). If no processor is available, emulation techniques are applied. For example, with In-Circuit Emulation (ICE) the microprocessor can be replaced by a tester-connected plug and the tester takes control of the signal processing; the tester also provides a debug facility for the processor's input/output signals. The same can be done for ROM emulation. In other cases the board test code may share the on-board RAM or ROM code space.

So, in general, a functional self-test comprises an on-board (or emulated) processor providing the stimuli, observing the responses, comparing them with the expected values and reacting on errors. Normally the more primitive functions for the microprocessor's environment (kernel, memories and bus functions) are verified first. Once established as error free, the other modules/subsystems are tested sequentially.

Concurrent sampling during board FST means the following. At predetermined instants and intervals during the continuous execution of the FST, data are strobed by the Boundary-Scan cells at the chip's I/O pins. The sampled data is shifted into a compressor while the microprocessor continues to execute code. This action is called a sample-scan cycle. These cycles may run continuously in a free-running style or be triggered by the processor. In either way, a complete *sample-scan cycle test* consists of an initialization, multiple sample-scan cycles and a contents/signature verification. Since the read signature depends on the sampled data input sequence and the FST code segments may be run in any order, each sample-scan cycle test must be completed within one code segment.

Features

Chip I/O values of interest can be sampled safely only if the sample clock(s) can be synchronized to the board system clock(s). Usually all Boundary-Scan chips receive the same copy of the sample clock. In addition, the test data sampled by the parallel Boundary-Scan cell must precede the sample clock by at least the cell set-up time and persist for at least the hold time of the cell.

Sampled signals to be compressed must not have indeterminate ('X') value because they will corrupt the compressor response. Therefore, these bits are masked, giving a fixed value before they are shifted into the compressor. Generation of the mask data requires circuit knowledge and analysis to predict which sampled I/O data will have known values. If there are on a board N Boundary-Scan registers each having an assumed equal length of M, then N·M bits must be stored in the on-board memory tester. The mask data are stored in a N·M Mask RAM (N words of M bits). For simplicity of the signal compression, a signature analysis with a Multiple-Input Signature Analysis Register (MISAR) could be used.

Figure 3-1 shows a possible board implementation including a clock synchronization unit, the mask RAM and the MISAR. The nets 1, 2, 3 and 4 allow various sample-scan cycles.

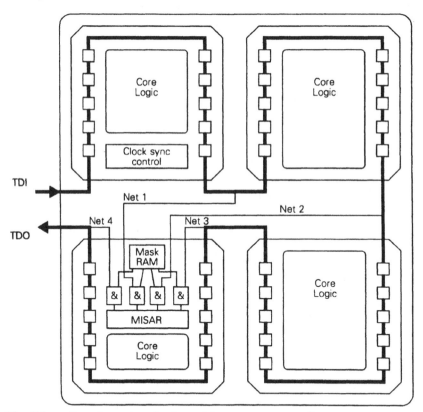

Fig. 3-1   Board structure for test with concurrent sampling

Costs and Advantages

For implementation of this test method only a relatively small chip area is required. Suppose a board has 10 ICs and the sample-scan is needed for 200 I/O pins per chip. A 16-bit MISAR is required and a 2 kB Mask RAM (if needed) would suffice. In a 60k circuit CMOS ASIC the area overhead occupies about 3% or even 1% if the Mask RAM is not needed. At board level (10 ICs) these figures are reduced to 0.3% and 0.1% respectively.

Compared with functional testing alone, this method improves the effectiveness of both defect detection and diagnosis. To avoid explicit bit by bit monitoring of the I/O signals, the responses are compressed, even during normal FST operations. It should be noted that concurrent sampling augments the FST capabilities, it does not replace FST. However, the FST code volume is reduced due to the increased effectiveness of concurrent sampling.

The Boundary-Scan Master

In essence the Boundary-Scan Master (BSM) is an intelligent parallel-serial protocol conversion device to provide a board-level built-in self test (BIST) requiring minimal hardware and software development effort. It is an enhancement of a test strategy which is based on the IEEE Std 1149.1 invented by AT&T Bell Laboratories [6]. The enhancements address a number of items. For example, to meet the throughput requirements when large amounts of code are involved, a dedicated bus master providing the parallel-serial interface between the board tester and the TAP is useful. To vary the execution speed, stimuli can be produced with a hardware pattern generator and the responses can be compressed into a signature. To avoid corruption of the signature analyzer, the signature requests the masking of undefined bits.

Features

The following description refers to figure 3-2. The BSM front end contains a generic, microprocessor controlled interface with a data and a control bus. The other end provides the IEEE Std 1149.1 TAP control signals, i.e. TDO, TDI, TMS and CLK.

Within the BSM the ATPG (Automatic Test Pattern Generator) comprises six control/status registers providing the following functions.

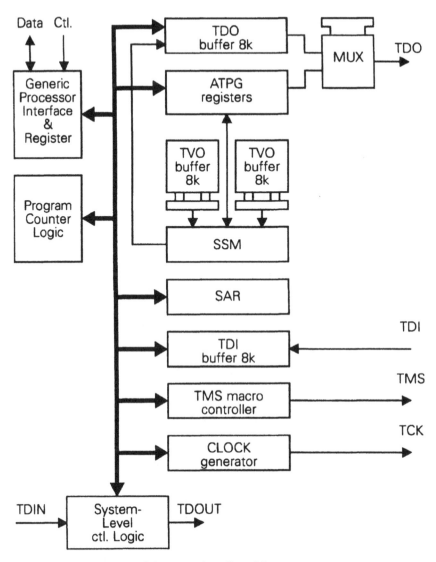

Fig. 3-2   Block diagram of the Boundary-Scan Master

- Test resource control, determining the source of the tests applied to the board units under test and the destination of the responses.

- Selection of one of the four possible vector sequences that the ATPG is able to generate:
  - pseudo-random
  - walking sequence (walking ones and zeros)
  - counting sequence (up or down)
  - constant output (ones or zeros)

- Control of the program counter logic. In particular the number of parallel test vectors to be applied to the unit under test.

- Control of the vector length for each scan path. For example, the length of the Boundary-Scan Register.

- STorage of the counts of the stimuli cells in the scan path. If the test is an interconnect test, then this register stores the total number of nets that are to be tested.

- Containing the signature analysis register (SAR) and a pseudo-random pattern generator, both 32 bits. Both seeds are programmable, although the pattern generator is write-only.

The next block is the Scan Sequence Modifier (SSM) which modifies a test sequence to ensure that no conflicts will occur before it is passed to the Boundary-Scan chain. The SSM also enhances the efficiency of the pattern generator (ATPG) and it controls the SAR while it compresses the selected values.

The SSM interacts with the ATPG (generator mode) and the SAR (deterministic mode). In either case, the SAR is the final destination of the TDI signals. In the generator mode the two 8 kB test vector buffers (TVI and TVO) hold the structure mapping of the scan path. There is a one to one mapping between the N locations of this buffer and the N Boundary-Scan cells in each of the M chains of the unit under test. The TVI and TVO memories are initialized by the board tester. TVI identifies the control or data cells that must be held constant while TDO supplies the data for these cells and identifies the stimuli/response cells.

A '1' in a location of TVI identifies a control cell at the corresponding location of the BS chain. The value in the same TVO location contains the data that must be shifted into that cell.

A '0' in a location of TVI implies that the cell is either a stimulus or a response cell, determined by the value of the TVO cell. It will be a stimulus cell (hence the ATPG is enabled to supply test data for the cell) if TVO contains the value '1'. It will be a response cell if TVO contains the value '0'.

When the SSM is in its deterministic mode with the SAR as TDI destination, then the vectors stored in the TVO buffer are assumed to be correct, i.e. no conflicts should occur in the SAR. In this mode the TVI information is used to enable the SAR; a '1' in a location of TVI enables the SAR and the information of the cell in that location is compressed.

The TMS Macro Controller synchronizes the operations of the ATPG and the SSM. The ATPG starts generating serialized algorithmic or pseudo-random vectors from the TDO buffer as soon as the TMS controller signals the ATPG that the TAP Controller has entered its shift state.

Costs and Advantages

Board-level BIST targeted towards automated generation of test stimuli and compressing the responses to a signature, is provided at the cost of an extra IC. The Boundary-Scan Master is capable of running a five-step test: the BSM self-test, the Boundary-Scan path integrity test, the board interconnect test, activating the device (IC) BIST and perform a cluster test. Automated pattern generation keeps the programming costs to a minimum because the board's design information (the netlist and a BSDL description) is available.

Memory Board Testing

Another special device providing board-level BIST is dedicated to memory boards. Like the Boundary-Scan Master it is offered by AT&T Bell Laboratories [7]. This is a VLSI design called the Memory Control, Error Regulation and Test (MCERT) chip. It integrates these test functions for application in board-level memory arrays. Moreover, programmable memory test algorithms can be invoked by the system user. Its Boundary-Scan interface is compatible with the IEEE Std 1149.1.

Traditionally memory boards are tested by software algorithms as part of a system test. While the assisted built-in self test (BIST) is already available for ASICs (including their embedded memory), such a facility for a board containing a number of RAMs is not widely spread. Such boards may contain over 100 MB of memory and testing through a sequence write/read operations marching through all locations of the memory can take a considerable time. If, for example, such a test is to be repeated 14 times, then testing such a loaded memory board by software algorithms may take some 10 to 20 minutes.

Features

The MCERT chip is made up of two distinct sections: a data path section and a control section. Each section contains 32 bits wide registers.

The control section of the chip has 17 registers, classified as command register, status register, march test registers, bank address boundary registers etc. The internal registers must be programmed before the chip may be used for memory access operations. These registers are accessed as programmed I/O operations through the system address and data ports of the chip. Main memory access requests and the programmed I/O operations are both handled in the same way.

The chip accepts 36 bits of data (32 data bits and 4 parity bits) and 34 bits of address (30 address bits and 4 parity bits) from a source requesting access to the memory or to the internal registers. It provides 32 data bits and 7 check bits to the memory array.

The MCERT chip is designed to execute user defined memory test algorithms. Four march test registers in the control section contain the test algorithms. The programmable test algorithm can detect a large class of functional faults in the memory array, such as stuck-at and coupling faults.

The data path section contains a 39 bits wide test data generator, which can deliver two pairs of test data: {<0000...00>, <1111...11>} and {<0101...01>, <1010...10>}. During memory test, the check bit generator is disabled and the entire 39 bits wide word of memory is treated as data.

The march test registers allow user defined test algorithms for memory marching. A march test is made up of a finite sequence of march elements, which in itself are a finite sequence of march operations. The operations pass every cell in a memory in either increasing or decreasing address order. The MCERT allows also programmable march elements. These elements may contain from one to seven operations. The march test registers are 32 bits wide. The user programmable march elements provide a maximum flexibility in designing a test algorithm for a given memory array fault model. The following test features are available in the memory test mode.

•   A memory test may be done on selected banks of memory.

•   Tests may be stopped on the occurrence of the first memory error and restarted from the last failing location.

•   A failing location is identified by the bank that failed, the address that failed and the data bit that failed.

- At the end of a successful memory test, all the memory locations are automatically initialized with the correct check bit values.

The MCERT chip supports the IEEE Std 1149.1 Boundary-Scan. So this feature can be used, amongst others, to execute a BIST (memory test) on a board. In this way the MCERT enhances the Boundary-Scan test. Figure 3-3 shows the MCERT's 1149.1 compatible Boundary-Scan interface (compare also figure 2-1).

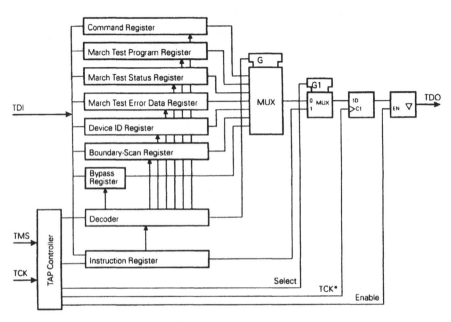

Fig. 3-3   Boundary-Scan interface of the MCERT chip

Testing a pure memory board in the factory implies a functional test through the board connectors at the highest frequency supported by the board tester. This functional test is a standard march test for memory, writing in each location a logical '0' followed by a read operation, and repeat the same action for a logical '1'. Though this is a simple test, it has still to be developed, the execution is lengthy and a (computer aided) tester is required.

Boards containing a mixture of logic and memory devices must be tested by a complex and expensive tester. These tests are really expensive to develop and they usually run at a relatively low clock rate. The MCERT chip allows the command register and the march test registers in its control section to be programmed through the 1149.1 Boundary-Scan interface. A march test status register is also accessible through this interface. In the factory a default march test may be initiated by just powering up the board. At the end of the test, the march test status register may be scanned out in the Boundary-Scan mode to read the result of the memory test. If a subsequent memory test is to be performed, the march test registers may be reprogrammed, again through the Boundary-Scan interface.

Costs and Advantages

By using one VLSI chip on a board, the factory costs are significantly reduced through both less test development and test equipment costs. The throughput is also greatly increased due to reduced test times. Memory tests have shown 20 to 30 times shorter test times by introducing the MCERT chip. For example, a marching 1 and a marching 0 test for highly loaded memory board may take 520 seconds with a software run test but the run by MCERT takes only 18.6 seconds. Finally, the MCERT provided BIST facility is accessible by the system user as well as by the factory personnel.

In chapter 5 we describe how the interconnections of a memory cluster on a PCB can be tested applying Boundary-Scan Test. For that purpose adapted test algorithms have been developed, resulting in interconnection test times of less than a millisecond.

SYSTEM-LEVEL TEST SUPPORT

In the same way that the IEEE Std 1149.1 TAP and Boundary-Scan Architecture can be extended from IC level to board level by connecting the IC TAPs, a system wide 1149.1 compatible test methodology may be obtained by utilizing the board-level TAPs at the system backplane. Since various industry groups are interested in this application, it is obvious that there is a desire for an additional test standard. A proposal for such a standard is described in the IEEE P1149.5 document [8], which will eventually become a formal standard.

Embedded Go/No-Go Test With BST

Most systems consist a Power-On-Self-Test (POST) unit which checks on failures at power-up time and resets/initializes all digital system logic. Such a self test improves the reliability of the system, which in turn offers higher quality to the customer. This POST can be extended for testing the interconnections on a PCB that complies with the IEEE Std 1149.1 for Boundary-Scan testing. It should be mentioned that POST performs only a go/no-go test, thus the same applies to the BST board, i.e. no detailed diagnostics down to IC pin level is required. A pass/fail message for interconnect test of the BST board suffices.

Moreover, no test speed requirements for the Boundary-Scan go/no-go test are required. Therefore a standard I/O system port of the present host CPU can be used to perform the test. Since the available system logic is used no extra hardware is required and thus a 'zero-cost solution' is obtained here.

The configuration for this test is depicted in figure 3-4.

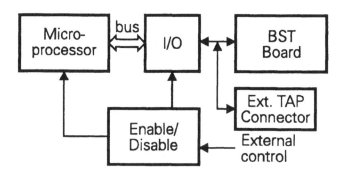

Fig. 3-4   Power-On-Self-Test for a BST board

The POST provides an external signal (static) to the enable/disable logic for the I/O bus in order to prevent bus conflicts. The external TAP connector is usually present for manufacturing and field service reasons. In case of a failure it permits a dedicated Boundary-Scan tester to be connected. This will diagnose the errors and display an accurate report, specifying exactly the device(s) and the pins at which the fault occurs.

Both the embedded go/no-go Boundary-Scan test and the external connected BS tester comply with the IEEE Std 1149.1, guaranteeing the compatibility between the two test methods. This in turn means that every tester that claims IEEE Std 1149.1 compatibility can be connected to the external TAP connector.

To perform the tests using the embedded CPU serialized test vectors are required. An extremely simple routine can then be used to shift the test vectors into the Boundary-Scan chains on the PCB and compare the responses to the expected values in the POST logic.

System Backplane Test Bus

Typically, at board-level, the Test Access Ports (TAPs) of all 1149.1 Boundary-Scan compatible ICs are daisy-chained to make a complete board test path, also called a test ring. This is made available through the TAP, usually at the board's edge connector. Creating a system-level test ring requires that the system design includes an appropriate way to make all on-board TAP signals available to the system backplane. This can be realized in various ways.

- Bringing out the TAPs of all boards to the system backplane and using a suitable multiplexing scheme that selects one or more board-level TAPs. This is easy to design but will introduce a large number of wires to the backplane. Usually one will refrain from such an application for feasibility and economical reasons.

- Daisy-chaining the board TDI-TDO pins to form a system-level test ring on the backplane. A huge drawback of such a method is that when the system is left with an empty slot for a board, the system test ring is broken in arbitrary positions. This might be cured with some physical connections but it remains clumsy.

- Using a dedicated chip to link the board-level TAPs and test rings to a structural test and maintenance bus on the system backplane. This uses special chips on board-level designs, like those described in the following sections. The disadvantages of the previous two possibilities can be avoided in this scenario.

An example of the last proposition is given by Digital Equipment Corporation [9] and is summarized here.

The architecture comprises the following basic elements, see figure 3-5.

1. The system backplane, providing a electro-mechanical construction that holds the slots into which the printed circuit boards (PCBs) are put. Each slot is numbered (1...n) and is identified by its Slot-ID.

2. The backplane test bus consists of the 1149.1 TDI, TMS, TCK and TDO connections. The connection for the optional TRST* signal may be incorporated as well. Note that the TDI and TDO signal paths are not daisy-chained but are bussed.

3. The local test ring, which is the daisy-chained connection of the on-board TAPs of the ICs compliant with the 1149.1. For the time being one single test ring per board is assumed.

4. The BT-Link unit, a Backplane Test Bus Link providing the interface between the local test ring (on-board logic) and the system backplane.

5. The Test and Maintenance Master (T&M Master) is a control unit for the backplane test procedures. The T&M Master may reside on one of the PCBs or

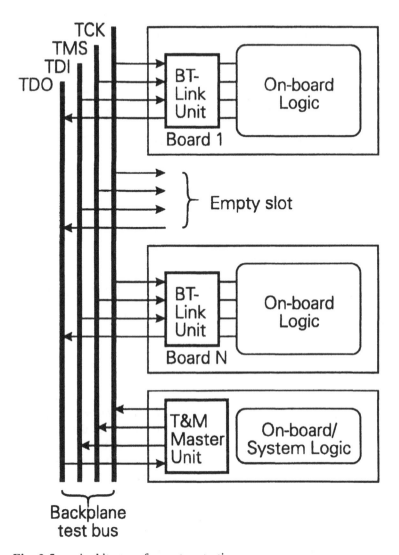

Fig. 3-5   Architecture for system testing

on a system board, e.g. the power supply board. For explanatory purposes we will assume one single Master in the system.

To explain the architecture's properties the interactions of the two most dominant units (BT-Link and the T&M Master) are discussed here.

BT-Link Unit

Figure 3-6 shows the BT-Link unit. The T&M Master unit can move the BT-Link TAP Controller through all its states.

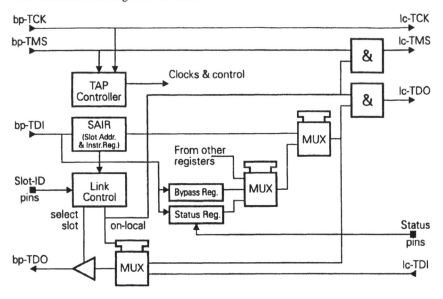

Fig. 3-6   Blockdiagram of the BT-Link unit

On the left hand side the backplane test bus (TAP) is provided together with the Slot-ID pins, while the BT-Link's output contains the local TAP to the on-board logic and the optional Status Pins. The slot-ID provides the identification of the board. For a system backplane of N slots, the size of the slot-ID is at least $\lceil \log (N+2) \rceil$ bits. The delimiters '$\lceil$' and '$\rceil$' denote that the next higher whole number is to be taken if the calculated expression between them does not result in a whole number. The slot-ID codes of all zeros and all ones are reserved, so a five bits slot supports 30 slot addresses.

The Slot Address and Instruction Register (SAIR) is similar to the 1149.1 Instruction Register and hence conforms to the standard requirements. It receives the slot address and it holds the 1149.1 instructions. It supplies two sets of parallel outputs. One set is taken from the parallel output latches and serves as normal 1149.1 instruction. The other set is taken from its shift register stage and is interpreted as the slot address during the link state.

The BT-Link is forced to enter the Link-Wait state when its TAP Controller enters the Test-Logic-Reset state. It remains in this state until the TAP Controller passes through the Update-IR state (see figure 3-7).

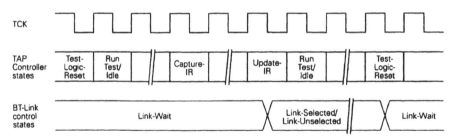

Fig. 3-7   Timing of BT-Link control logic

In the Link-Wait state the BT-Link is waiting for a slot access request from the T&M Master. The bp_TDO driver is then inactive or tristated. The updates to the IR are also inhibited, meaning that the IR retains either its BYPASS or IDCODE state. The data now received at the bp_TDI pin is interpreted as the slot address. The slot address in the SAIR shift register stage is compared with the Slot-ID when the TAP Controller enters the Update-IR state; when the TAP controller exits from the Update-IR state, the BT-Link leaves the Link-Wait state.

When the Link-Wait state is left, the BT-Link enters either the Link-Selected or the Link-Unselected state, depending on whether or not the received slot address matches the Slot-ID.

When the Link-Selected state is entered, the bp_TDO is enabled during the Shift-IR and the Shift-DR states. The updates to the IR are also enabled. One of the instructions supported on the BT-Link connects the on-board logic (test ring) to the backplane test bus. This situation lasts until the TAP Controller returns to the Test-Logic-Reset state at which the BT-Link returns to the Link-Wait state.

When the Link-Unselected state is entered, the bp_TDO driver remains disabled or tristated. The TAP Controller continues to follow the T&M Master and updates to the IR remain inhibited, meaning that the IR retains either its BYPASS or IDCODE state. This situation lasts until the TAP Controller returns to the Test-Logic-Reset state. Then the BT-Link returns to the Link-Wait state.

Note in figure 3-7 that the changes in the BT-Link Control logic occur on the falling edges (negative slopes) of the TCK.

BT-Link Instructions

Two instructions for the link are always needed besides some optional public instructions.

The BYPASS instruction is needed and its opcodes consist of all ones. Obviously, it selects the DT-Link's BYPASS Register.

The other instruction needed is the ON_LOCAL, that enables the connection between the on-board logic (test ring) and the backplane test bus. The connection remains until the optional OFF_LOCAL instruction is entered or until the BT_Link TAP Controller is forced by the T&M Master into its Test-Logic-Reset state.

The OFF_LOCAL is a user instruction that disables the on-board logic's TAP and so disconnects the on-board logic from the backplane test bus. The instruction forces a high (1) on the lc_TMS and lc_TDO pins. Consequently the on-board logic TAP controllers are returned to their Test-Logic-Reset state in three TCK cycles.

The optional SEL_STATUS is provided if the optional Status Register and Status Pins are implemented. This instruction selects the BT-Link's Status Register for scan operations, allowing the T&M Master to read the status flags that may be set by the on-board logic circuitry via the Status Pins.

Finally, the BT-Link module is powered-up in compliance with the IEEE Std 1149.1: the TAP Controller is forced into its Test-Logic-Reset state and the BT-Link control logic is forced to enter the Link-Wait state.

T&M Master Control

When the system needs access to a specific board, the T&M Master unit shifts the requested slot address simultaneously into all BT-Links. During the Update-IR state, each BT-Link unit compares the received slot address with its own Slot-ID. Only the BT-Link that successfully matches its Slot-ID with the slot address enters the Link-Selected state. All other BT-Link units enter the Link-Unselected state. As described above, this situation remains until the BT-Link TAP Controllers are returned to the Test-Logic-Rest state by the T&M Master. If none of the BT-Links match the broadcast slot address, all subsequently received data are ignored by all BT-Links until the T&M Master returns the BT-Link TAP Controllers to the Test-Logic-Reset state. This situation is likely to arise when the addressed slot is empty.

The selected BT-Link enables its TDO driver and connects with the backplane test bus. To establish connection to the on-board logic (test ring), the T&M Master utilizes the ON_LOCAL instruction. As soon as the ON_LOCAL instruction becomes current in the Update-IR state, the on-board TAP is enabled and further protocols received from the backplane test bus are also transmitted to the on-board test ring. Now the T&M Master can operate any IEEE Std 1149.1 instruction by shifting in the appropriate code.

Note that, since the BT-Link unit and the on-board logic test ring operate in series, the BT-Link TAP Controller and the on-board TAP Controller must operate synchronously. As a matter of detail, this constraint is inherently met when the T&M Master leaves the Update-IR state that established the ON_LOCAL instruction and returns to the Run-Test/Idle state.

When the test protocol is completed, the access to the on-board test ring is turned off either by shifting the OFF_LOCAL instruction or by returning the TAP Controller to its Test-Logic-Reset state with CLK pulses. The latter also forces the BT-Link unit to enter the Link-Wait state. The system is ready for the following test operation.

Costs and Advantages

The architecture uses a backplane test bus and a BT-Link unit. The latter is of a rather simple design and may be composed of readily available standard components. As such it should not be an expensive module. It could even be implemented in ICs which are already 1149.1 compatible, which again should be an inexpensive extension of the host IC design, as already indicated.

As stated before, an IEEE P1149.5 document is issued. However, the proposed standard architecture for system test is more sophisticated than the backplane system described here. For example the IEEE P1149.5 document proposes a structured communication protocol, a degree of fault detection and some fault tolerance, to name but a few items.

To summarize, a few advantages of the described system test which is based on an 1149.1 enhancement are:

- It uses only one chip-to-system test access protocol. Hence, the test control design as well as the test software is simpler than as proposed in the IEEE P1149.5 document.

- With the optional Status Pins and Status Registers a back-door interrogation of the on-board status is possible, without invoking the on-board's test ring.

- The system can be enhanced so that the BT-Link unit contains a module's identification, like the component's Device-ID.

ON CHIP PROVISIONS

The definition of an IC varies widely. It may be an individual die (chip), a Wafer Scale Integration (WSI) design, consisting of multiple dies which are not broken in individual chips and which are interconnected on a monolithic wafer or possibly ever a Multi Chip Module (MCM) as an IC.

Various solutions to introducing BST into ICs are given in this section, together with their (possible) trade-offs.

BST on WSI designs

The efforts of implementing of the IEEE Std 1149.1 Boundary-Scan Test (BST) interface on an IC should always be counterbalanced with the expected benefits. On one hand there are the improved testability and the reduced I/O area needed to support the full wafer probing, on the other the loss of area yield due to the needed BST circuit area. Moreover, extra provisions may be needed to allow for tests which are not particularly suited to Boundary-Scan tests, for example an at-speed (self-)test of the core logic.

The yield of a Wafer Scale Integration (WSI) design is limited by the number and locations of defects on the wafer. Before a working system can be configured, the good cells must be identified. With wafer probe testing full size I/O pads and large output drivers to drive the capacitance introduced by the probes and the input circuitry of the tester are needed. See figure 3-8.

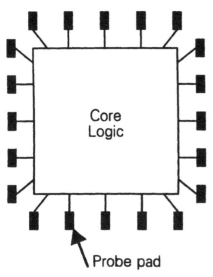

Fig. 3-8   Chip arrangement for full wafer probe testing

The driving of such buffers with its associated power consumption is not needed for inter-cell signal connections, as provided by the Boundary-Scan serial test interface. See figure 3-9.

Scan testing is even more profitable on wafer system (VLSI) level. In order to increase the wafer yield, an IC design and fabrication technology is developed which tolerates defects. Therefore redundant components are introduced, which can be used to circumvent the defect paths. The technique to restructure a VLSI is to use a laser beam to configure the correct wafer level interconnections. Connections can be established and deleted [10].

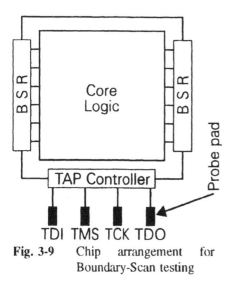

Fig. 3-9   Chip   arrangement   for
Boundary-Scan testing

Other methods include ion beam and laser enhanced deposition technologies.

Next, in order to test the individual chips prior to reconstructing the system at wafer level, a full wafer probe test technology would need probe pads around every single cell (chip). However, after configuring the VLSI system, a lot of these pads have become superfluous because only the pads needed for wire bonding on package (VLSI) level remain functional. Here again the Boundary-Scan test technology is advantageous because the 1149.1 allows a system level test of all chips, provided that all chip TAP connections are addressed by the system test bus. One more restriction to the Boundary-Scan application on chip level should be noted here. If a cell (chip) has a very small size, than the implementation of a Boundary-Scan Register would result in too much overhead of wafer area. This restriction together with the at-speed chip test capabilities before, during and after a wafer restructuring determine whether or not BST can be applied. Studies and workbenches have been carried out [11] to indicate the limits for the application of BST in the WSI design.

BST on Multichip Modules

Multi Chip Modules (MCMs) are modules built up of dies on top of several dielectric and metallization layers supported by a ceramic multi-layer substrate. These modules are so big, with functionality comparable to a loaded PCB, that they are only accessible via their input and output pins. So the testing of these modules becomes a problem. For example, due to its price, the MCMs do not lend themselves to a type of go-no-go testing, where modules that do no pass are thrown away. Therefore, to obtain a commercially viable test method for MCMs, both their architecture and the testability features have to be determined.

Modules this big are never tested for the first time in their final state. It is of course sensible to first test the constituting components, the ICs and the layers. Thus before setting up a MCM test architecture, it is good to ensure that:

• the components (ICs) have been tested before,

• the component's driving output and input current levels are within specifications,

- the dielectric and metallization layers are properly produced and

- the substrate with its interconnects have been tested.

Having settled these points, attention can be paid to the MCM itself. Due to the very high density packaging it must be stated that Boundary-Scan testing is the only applicable method for MCMs. Next, the module's architecture is to be considered in connection to testing. To that goal investigations [12] have revealed guidelines which are summarized here.

- It can be stated that the larger the size of an interconnect feature, the lower the probability is that a defect in it will adversely affect the circuit performance. Therefore, the interconnect geometry should be designed as large as is feasible. The same applies to the ASIC's feature size.

- ASIC probe pad drivers contribute to the signal propagation delay due to their inherent capacitances. Designing close to the driver's output current margins may result in undetected propagation delay defects. Therefore, ASIC pad drivers should be designed with a current safety margin of at least 150%.

- ICs having a 'typical' timing spec may lead to overall timing outside the designed range if the signal paths of several ICs are put in series. Therefore, system timings should be such that they function at worst-case specification of the ICs.

- A die may be placed on the wrong place in a circuit. Moreover, the handling and mounting of bare dies may damage them. Since only the Boundary-Scan test can be applied to test the MCM module-wide later on in the process, all ASICS should obey the IEEE Std 1149.1 and, additionally, have the following optional instructions available:
  - IDCODE and USERCODE if the device function is alterable,
  - RUNBIST and/or INTEST.
  Also, all 3-state and bidirectional outputs should be under control of the Boundary-Scan registers to facilitate EXTEST measurements.

- In order to cope with the fact that not all ICs on the MCM are designed with a Boundary-Scan facility, at least all Module I/O pins should be connected to a Boundary-Scan register cell. This precaution also significantly eases the verification of module connections to the outside circuitry once it is mounted onto a printed circuit board.

- Full test of on-module memory should be provided by some on-module logic, such as a memory controller, which itself can be driven by IEEE Std 1149.1 controllable circuitry. Due to its nature, on-module memory testing is a special technique, which is even more valid for dynamic memories. For the latter type of memory the leakage current, gate oxide problems etc. affect its contents. Since memory devices show very tight timing requirements, relatively few memory devices are expected to support the IEEE Std 1149.1.

These guidelines are not very difficult to follow and should be applied to MCM designs wherever possible, inherently ensuring the quality of an MCM.

Given the above guidelines, the test strategy for MCMs is as good as defined and can be summarized as follows.

- Test the connections to the MCM I/O pins. This test is of utmost importance because the module I/O pins provide the only access to the tester. Following the guideline to have a Boundary-Scan cell at each Module I/O pin will greatly simplify this test.

- Test the IEEE Std 1149.1 TAP and BS infra-structure integrity and the IDCODES. It is essential to first check, diagnose and correct these items before any other test is performed. In a later subsection this subject is treated separately.

- Test the chip interconnections on opens and shorts.

- Test the on-module's bus integrity. This check verifies that no opens exist at the output and/or bidirectional pins of those ICs which are bussed together.

- Test the on-module device integrity. Dies may be damaged during the module's fabrication process. Therefore tests like RUNBIST and INTEST verify the functionality of the individual devices.

Note that all these tests are performed with Boundary-Scan testing, preferably in the sequence as given here. A failure in any part of the Boundary-Scan chain must be repaired before further testing of the MCM is possible.

BST on MOS Designs

Preparing a device to meet the IEEE Std 1149.1 test standard will always require a couple of decisions to be taken. Adding the IEEE Std 1149.1 testability to a circuit design must be weighted against the major challenges of minimizing both silicon area and the manpower spent on the design and layout.

For a number of VLSI CMOS microprocessors Motorola has introduced the IEEE Std 1149.1 in their designs [13]. The efforts taken are described, together with the problems encountered, the solutions found and the lessons learned. In this subsection some of the conclusions are summarized to give another view of the provisions to be made when implementing the IEEE Std 1149.1 into a hardware design.

- *Instruction Decoding*
  It is suggested that all instructions be fully decoded and documented as a unique opcode. Unused opcodes can be documented as "reserved" and decoded to select the BYPASS register. This practice is better than to document unused codes as "don't care" bits, which will forfeit the manufacturer's flexibility of having unused opcodes and subject the customers to change their application if instructions are added.

- *TAP Controller*
  Extra effort may be required to imply circuitry for an asynchronous reset of the TAP controller. The IEEE Std 1149.1 prescribes that all TAP controller state actions occur at a rising edge (positive slope) of the TCK, whereas actions of the test logic may occur at either the rising or falling edge of the TCK. Now suppose that an on-chip logic block requires an asynchronous reset at the same instant that the TAP controller is reset. Though not difficult to implement, extra effort (chip area and design power) is needed to make the design functional.

- *Boundary-Scan Cells*
  A design philosophy seeking for a minimum design effort leads to a modular approach to the cell layout. So, if for example many bidirectional pins are required in a design, one might decide to pair each pin with a bidirectional control cell. It turned out that this implementation facilitated increased flexibility during board-level tests, despite the fact that such a Boundary-Scan cell is much larger than a conventional, comparable output cell.

- *INTEST Instruction*
  The biggest problem for this optional instruction is to find a manner that provides utility in the stimuli and responses which can be practically generated by the user. Given a VLSI design, a cycle-by-cycle bus behaviour appears to be only loosely correlated with the many lines of code in an assembly language listing. For example, there may be no common cycle period between the CPU and (one of) the different peripheral modules. It was therefore decided NOT to support this optional instruction.

- *SAMPLE/PRELOAD Instruction*
  The SAMPLE instruction uses the Boundary-Scan register to provide a snapshot of the state of the IC pins during normal system operation. A typical SAMPLE problem is that there is no instant that all the pins are in a logically valid state. Different outputs are issued at different edges (slopes) of the system clock. A synchronization mechanism for the various clocks, therefore, is necessary.

  Another problem is the timing skew at various points in the design.

1. There can be a timing skew in the phase difference of TCK pulses as measured at one device and at the Boundary-Scan cell with which it is operating.

2. Within the device, there is a timing skew between the system and the test logic operation. Differences exist in times between applying edges to the system clock and TCK, while responses are observed in the system logic and test logic respectively. This skew even varies with the temperature.

3. Timing skew exists also in the IEEE Std 1149.1 Boundary-Scan logic itself. Timing and control signals are routed from the TAP controller in *both* directions around the periphery of the chip (see also chapter 2). The final Boundary-Scan performance is determined by the cells that are physically most remote to the TAP controller.

Despite these signal validity and skewing problems the SAMPLE utility is considered to be useful for the end users.

While implementing the IEEE Std 1149.1 Boundary-Scan provisions into their VLSI designs, Motorola has taken care that non-1149.1 users are not burdened by the inclusion of these features.

As an example figure 3-10 depicts a MOS design into which the IEEE Std 1149.1 circuitry is highly integrated with the system circuitry of a pad driver.

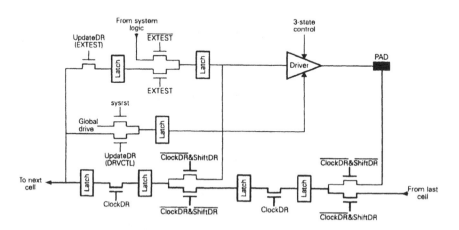

Fig. 3-10   Integrated BS cell for Motorola 68040 address pins

Digital Bus Monitor

The ever increasing speed and complexity of IC designs will force test methods to become faster and faster, for both internal IC testing and testing at wired board level. This has led to a new test approach, allowing board level functional testing to be performed at-speed by an on-chip test-logic. The IEEE Std 1149.1 Boundary-Scan test standard provides the instructions by which the IC on a board can be placed off-line: the SAMPLE/PRELOAD and the EXTEST (see chapter 2). With respect to the speed of Boundary-Scan testing, a number of *potential* problems can be expected [14].

- In order to fetch stable data from the boundary of the target IC, the scan clock (TCK) should be synchronized to the rate at which data are traversing through the boundary. If the IC has its own system clock it can be substituted for the scan clock by adding on-chip switching circuitry.

- A system clock which is used as scan clock, may be too fast for the IEEE Std 1149.1 TAP and the SAMPLE/PRELOAD instruction can not be executed.

- In a typical board design not all ICs operate at the same system clock. Therefore it may be difficult to obtain valid data from all IC boundaries with one sample operation. It is also important to remember that the maximum data sample rate is determined by the maximum TCK rate of the slowest TAP Controller on the board design.

To circumvent these problems, additional board logic could be designed and implemented. Note that such board logic is to be added to each board that must be included in the system testing. But it can turn out that, with increasing operating speeds, board testing becomes impractical and very expensive.

A new approach to functionally test devices at-speed was invented by Texas Instruments Inc. [15]. A special IC has been designed for embedded, on-line at-speed testing. This design can not only be implemented in ICs but also on a multi-chip module and printed circuit boards. The IC design can be coupled to IEEE Std 1149.1 bus signals to provide an off-line method of non-intrusively monitoring the functional operation of the circuit. The design is called Digital Bus Monitor (DBM).

Features

The new method is a two step-approach: first enable an IC's boundary test logic to sample system data and secondly access the results for analysis. To that goal the IEEE Std 1149.1 Boundary-Scan test logic in the DBM is blended with an event qualification architecture. When the DBM is enabled via the serial input from the IEEE Std 1149.1 test bus, it is synchronized with the functional circuitry. Then data trace and/or data compaction is performed at-speed on the data flow and the trace data and/or signature collected can be accessed via the 1149.1 test bus for processing. The DBM of Texas Instrument Inc. (TI) supports all required 1149.1 Boundary-Scan test instructions as well as dedicated instructions supporting the data trace and signature analysis test operations. The latter operations can be controlled either by the 1149.1 test bus or by internal DBM control signals. Figure 3-11 depicts the architecture.

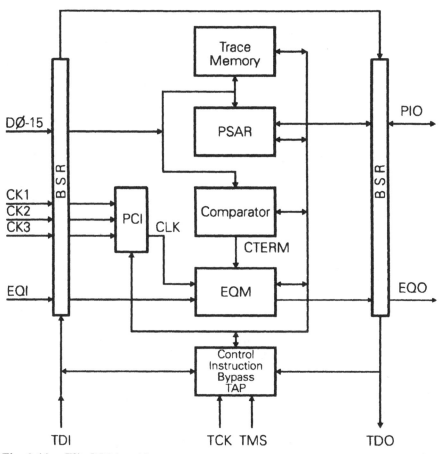

Fig. 3-11   TI's DBM architecture

The pins D0-15 make the 16-bits data input port from which the input data are fed to the 1Kx16 Trace Memory, the Parallel Signature Analysis Register (PSAR) and to a 16 bit Comparator for pattern detection. The output of the comparator, Compare Term (CTERM), is put into an Event Qualification Module (EQM) for at-speed test control, for which also the Event Qualification Input (EQI) is used. Its results are output as the Event Qualification Output (EQO) signal. PCI is the Programmable

Clock Interface which has control over the EQM via the CLK signal and which itself is controlled by the three clock input signals CK1, CK2 and CK3. All the circuit interactions are controlled by a control block, which also contains the 1149.1 Boundary-Scan circuitry such as the TAP Controller, BYPASS register, the Instruction Register and the data lines for TMS, TCK, TDI and TDO which are also connected to the Boundary-Scan Registers (BSR). PIO is a bidirectional Polynomial Input/Output signal pin of the PSAR.

The functions and features of the various circuit blocks are briefly described below.

The *Trace Memory* can receive and store data words at up to 40 MHz, coming from the 16-bit data bus. It also receives data in serial format from the TDI pin. Its control input comes from either the TAP or the EQM and the output goes to the other DBM blocks and in serial format to the TDO pin. The memory itself is a static RAM, 16-bits by 1024 locations.

The RAM has two modes of serial read/write access, the Boundary-Scan access mode and the Direct Memory Access (DMA). In the DMA mode the RAM's 16-bits Data Input/Output Register (DIOR) performs read operations in a parallel input/serial output mode to reduce access time. For the same reason DMA write operations are done in the serial input/parallel output mode. Calculations show time savings of up to 93% when comparing the DBM's memory access using the DMA mode over the 1149.1 Boundary-Scan mode, assuming a 5 MHz TCK frequency and a 40 MHz data rate. When disabled the memory maintains its contents in a powered down mode for minimum power consumption. The Trace Memory contains a self-test function to enable testing in both an IC production and field test environment.

The function of the *PSAR* (Parallel Signature Analysis Register) is to sample or compact data input to the DBM. The PSAR contains a bit masking circuitry and a Multiple Input Signature Register (MISR). The masking circuitry can be controlled to selectively mask one or more of the input data bits from being compacted in the MISR. This bit masking feature allows the MISR to operate as a parallel or serial input signature analyzer.

The PSAR receives the 16-bit data bus, and a serial input from the TDI pin. Its control signals come from either the TAP or the EQM. The selection of which control is used is determined by the instruction in the 1149.1 Boundary-Scan Instruction Register. In the latter case data input on the data bus is compacted into the MISR during each TCK while the TAP controller is in its Run-Test/Idle state. If the PSAR operations are controlled from the EQM, data input on the data bus is compacted into the MISR when the EQM is qualified to execute one of its eight test protocols.

After completion of each data compaction, the signature can be shifted out to the TDO pin. Output can also be coupled to a bidirectional pin (PIO) which allows cascading the PSARs of multiple DBMs.

The PSAR can receive and compact data words at up to 40 MHz.

The *Comparator* is used to detect patterns during data trace and compaction test operations controlled by the EQM or the TAP Controller. For that purpose the comparator comprises a 16-bit compare circuit and two 16x16 memories used to store the expected and the mask data patterns. The comparator receives the 16-bit data bus and serial input from the TDI pin. Its control signals come from either the TAP or the EQM. The detection result is output as a Compare Term (CTERM) signal which is fed to the EQM. The serial output is fed to the TDO pin.

The programmable *PCI* (Programmable Clock Interface) is used to minimize clock gating and selection operations. PCI inputs come from the CK1, CK2 and CK3 signal pins. It control signals come from the DBM's control block and the output is fed to the EQM. PCI clock inputs can be programmed to accept Motorola or Intel type read/write bus control without the need of any interfacing logic external to the DBM.

The *EQM* (Event Qualification Module) is a test control unit developed by TI to provide a standard method of enabling an IC's test circuitry during its normal operation. It can execute one of eight predefined test protocols. The protocol to be executed is input to the DBM controller's instruction register. The EQM receives its control from the TAP and operates synchronously to the PCI clock signal CLK. The parallel event input comes internally from the Comparator (CTERM) and externally from the EQI pin. Serial input data are provided by the TDI pin. In its output mode, the CTERM signal from the comparator can go to the EQO pin; the serial output data are fed to the TDO pin. The EQO signal can be used for triggering an external tester or as an input signal to a neighbouring IC as an EQI signal.

The protocols describe certain test actions. For example, on/during/after an Nth event do/start a test repeatedly (M times), or stop the test or pause testing during a number of clock pulses. Thus, before starting the EQM controlled test operation, the proper protocol must be selected, the protocol's control instruction must be defined and the number (M) of times must be determined for the test to be repeated.

As stated previously, the DBM can operate in two test modes, off-line under control of the 1149.1 Boundary-Scan TAP Controller and on-line controlled by the EQM. The off-line mode is primarily used to verify the operation of each component (IC) in a system (PCB, MCM etc.) and to check the integrity of the interconnections between the components. It is obvious that the DBM, where it complies fully with the 1149.1, supports all the mandatory 1149.1 instructions.

If an off-line test is successfully completed, there is a fair chance that the system will also operate at full speed. However, problems associated with the interactions between ICs may not be testable with an off-line test. Therefore an at-speed test is required to prove that all controlling software functions properly at the rated speed. In order to give an idea of the possibilities offered by the DBM protocols, some of the tests that can be performed at-speed are listed below:

• On-line testing during an event.

• On-line testing between two events.

• On-line testing in response to an event.

• On-line failure monitoring.

• On-line testing of system sub-circuits.

The advantages of applying a DBM are the same as for introducing Boundary-Scan testing. The initial costs of this test approach are repaid many fold in the factory and the field service.

Adjustable Scan Path Lengths

A way to improve scan test efficiency is to reduce the test vector length dynamically on those places and in those circumstances where it is appropriate. The basic idea behind this philosophy is that, considered statistically, the state of each cell in a scan register need not to be updated with every test vector. So if a dynamic partition of the scan register is possible, then only the relevant scan cells will be connected in a chain and need to be updated for a certain test. This in turn requires a shorter test vector for which less clock cycles are needed to shift data into the scan register cells.

For application of this technology it should be understood that a broad distribution in the frequency of usage of scan locations (pseudo inputs) must be present. Only then it makes sense to set up a dynamic partition of the scan path. However, justification is apparent in an automatically generated test, where an input is addressed only if specifically required. Important locations will be frequently addressed, while some locations may have marginal utility.

Another condition for this method should be a compliance with the IEEE Std 1149.1 for Boundary-Scan testing. This condition can be met easily as is indicated already in chapter 2, and the relevant figure is repeated here for convenience.

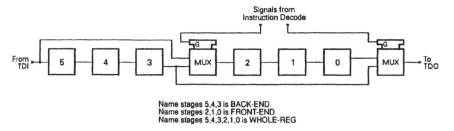

Name stages 5,4,3 is BACK-END
Name stages 2,1,0 is FRONT-END
Name stages 5,4,3,2,1,0 is WHOLE-REG

Fig. 3-12   Partitioning a scan test data register

With the example in figure 3-12 it may be sufficient to only change the values in the cells of the register "Back-end" while retaining the current values in register "Front-end". This requires a smaller number of shift cycles than a complete scan, which obviously, can still be performed when needed. Experiments at Racal-Redac Inc. [16] revealed that the benefit of dynamic scan path selection is optimized by placing the most frequently referenced scan locations in the first chain, "Back-end". Whether or not a scan is required is a function of the original vector set only and therefore is independent of any partitioning scheme.

The question remains how to determine the optimal partitioning of the scan register. For a set of stimulus vectors the partitioning should maximize the probability P_B that at least one cell of the mostly addressed "Back-end" must change while minimizing the probability P_F that at least one cell of "Front-end" must change. An algorithm to find the optimal partitioning could be the following.

- Compute for each scan cell the number of times its value changes for a given set of stimuli.

- Compute for each scan cell the probability that its value changes from one stimulus vector to the next.

- Sort the cells in descending probability and renumber them from 1 to B being the number of the mostly addressed "Back-end" cells and B+1 to T (of Total) being the numbers of remaining cells of "Front-end".

- For B=1 to T-1 calculate the above described probabilities P_B and P_F.

- Select the value of B which produces the smallest number of vectors.

The software for such a calculation is straight forward to write.

Two observations can be made with respect to this test methodology.

1. The above procedure can be used to estimate the optimal number of B. The accuracy of the result depends partly on how randomly the scan cell values change. Intuitively, one could expect a better accuracy the larger the set of stimulus vectors is. However, experiments on prototypes showed a rather low sensitivity to the test program length.

2. With a reduced amount of scan cells in the test path a negative impact on the fault coverage can be expected. Therefore, extra tests may have to be generated to achieve the same level of detection. But even when this is taken into account, a significant vector size reduction is achieved by partitioning the scan test path length.

Path-Delay Measurements

The design innovations described thus far in this chapter allow for at-speed measurements of logic circuitry. The next example shows another test innovation, in which the IEEE Std 1149.1 compatible Boundary-Scan cells used for interconnect faults (shorts, opens, stuck-ats, etc.) are adapted for at-speed measurements. Since these designs enable I/O delay measurements amongst others, they are referred to as path-delay measurements.

To explain the new design, two types of error are distinguished. First there are the static errors, which last permanently. Usually these errors are repeatable and can be detected independent of the operating environment, such as operating voltage or temperature. Within allowed specifications these errors can be measured at relatively low speeds of the test system. The other type of error is the timing dependency. Here all nodes of the circuitry may pass all desired signal transition tests but at a different speed than that specified. Usually test results show that a component is able to perform its intended function at a lower than the specified speed. But if certain changes occur at a much higher than the specified speed, then hold-time violations and signal delay times drastically reduce its performance and functionality.

One such design that copes with both the static and the dynamic failures in a digital integrated circuit is given by Hewlett Packard [17]. It goes without saying that the developed circuits comply with the IEEE Std 1149.1 for Boundary-Scan testing.

Internal Clock Signals

Only the functional clock signal of an IC can be used to capture the results of a path-delay measurement, regardless of how the test patterns are generated and applied for two reasons.

1. The internal clock delay, the skew and the at-speed set-up times of the various logic elements have to be considered as part of the internal path-delays. So their effects have to be measured and validated at clock speed.

2. Only the internal clock is a dedicated and reliable mechanism designed to enable the IC's internal logic elements (e.g. flip-flops) to operate accurately.

Special test and/or scan clocks, routed through the internal flip-flops, may exhibit greater delay/skew problems than the internal clock signal. Since test results are to be captured using the system clock, it follows that signal transitions which are part of the path-delay test pattern, must also be generated using the system clock. However, with scan techniques it is possible to initially load the internal flip-flops with a predetermined value. But the internal path-delay test pattern requires at least one system clock to create the output signal transitions.

The *basic idea* of the path-delay test philosophy presented here is the following. A Double-Strobe flip-flop (described next) has been designed to which two functional clock pulses (a "double strobe") can be applied. At the first clock pulse a signal transition is created on the input D which travels through the identified signal path. The flip-flop's output Z is clocked into a destination flip-flop on the next clock pulse. The construction of the Double-Strobe flip-flop allows the distinction between an initial and a final logic value at output Z. So the value captured in the destination flip-flop will have either the initial or the final value of output Z, depending upon whether or not two clock pulses are separated from one another by at least the propagation delay through the identified signal path. This property provides the method to repeatedly apply double strobes to the input while systematically bringing back the time between the two clock pulses. At a certain instant, the destination flip-flop is no longer capable of capturing the *final* value of the signal transition on the output Z. At this point the separation between the two clock pulses represents exactly the path-delay time.

Given this basic idea, some other issues must be considered. First the internal paths must be determined through which the path-delays must be measured. The selection of the subset of all possible signal paths in a VLSI must be done carefully to remain economical. A few scenarios are the following.

• Use the results of timing analyses, from where a number of the longest ('critical') paths can be selected and their path-delays measured.

- Include sufficient paths in the subset in order to cover all different types of primitive cells/circuit blocks so that errors due to the wrong characterization of these cells can be identified.

- Select those internal paths that pass through circuit nodes that are most heavily used.

One should next consider the generation of test patterns for the path-delay measurements, in which the test pattern itself must be identified. In particular, the static and transitional components of the signal must be distinguished. The static part component provides the sensitization of the transition path. A delay-test pattern is similar to a stuck-at test pattern except that in a delay test pattern the entire sensitization path is known.

The Double-Strobe Flip-Flop

This flip-flop can be implemented into an IC to allow path-delay measurements, as described above, without being constrained by dependencies among the inputs of the combinational circuit. For that purpose the internal flip-flops of the IC are designed so that they can be loaded with arbitrary *initial* values using the normal scan techniques as well as simultaneously storing a different *final* value in them. With the special test logic included in the chip it is possible to transfer the *final* value in place of the *initial* value under control of the regular system clock. If no signal transition is required at the output of a flip-flop then the same initial and final values are pre-loaded into it. However, if a signal transition is desired at the flip-flop's output then the initial and final values are loaded into that flip-flop so that its output changes from the initial to the final value on response to the system clock signals.

Figure 3-13 shows a simplified diagram of the double-strobe flip-flop or latch.

This circuit may be implemented using CMOS gate-array technology.

Features

The operation of this flip-flop is as follows. During normal system operation the scan clocks SI_CLK and SO_CLK are inactive and the control signals for the double_strobe DS and master-load ML are in their deasserted state. Now the flip-flop operates as a master/slave flip flop. The master latch is updated with input data D while the CLK is at its logic 0 level and the master latch is copied into the slave latch at the rising edge (positive slope) of the CLK pulse.

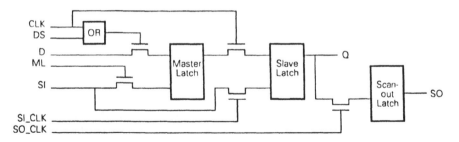

Fig. 3-13   Scannable Double-Strobe flip-flop

Regular scan operations can be performed using the SI_CLK to load the scan-in (SI) data into the slave latch and using SO_CLK to load data from the slave latch into the scan latch.

Next, to perform double-strobe testing it is necessary to assert both the DS and the ML signal. Then the master and slave latches are loaded with the same value utilizing the pair of SI_CLK and SO_CLK signals. Further, the master latch is designed such that, as long as the DS control signal is asserted, the master latch is isolated from all outside signal sources, so it retains its present value. This condition provides the possibility that different values can be scanned into the slave latch, by deasserting the ML signal and applying the SI_CLK and SO_CLK signals. Now, when such a scan cycle has been performed, the slave latch may contain the *initial* (inverse) value needed for the internal path-delay test pattern whereas the master latch contains the *final* value.

Path-delays are measured by applying two system clock pulses CLK to the flip-flop. The first CLK moves the inverse of the contents of the master latch into the slave latch. Inherent to the design of the double-strobe flip-flop, the first system clock pulse CLK applied also causes the DS signal to be deasserted. Thus when the next CLK pulse is applied, the double-strobe flip-flop operates in its regular master/slave mode and is able to capture test results.

When this test is successively repeated while systematically reducing the time between the two clock pulses, a point will be reached at which the system does not works properly. This time is the value of the path-delay time to be measured.

Though not designed for the purpose, the Double-Strobe flip-flop may provide an indication of the scan path flow-through time of a design (chip or board). If in figure 3-13 the scan clock signals SI_CLK and SO_CLK are held at a logic 1 level while the system clock is at logic 0 then both the slave latch and the scan-out latch are enabled. The SI signal may flow through the latches towards the SO output.

Thereby the SI signal is delayed by the circuitry in the latches. If in a scan design all or a number of those flip-flops are daisy chained, then the time delay of an SI signal applied to the input can be measured with a timing analyzer at the output of the chain. Such a delay may give a worst case/best case indication of the scan path flow-through time delay of the design under test.

Note that this measurement provides only an indication and should not be used as an accurate overall timing test.

The Double-Strobe flip-flop was designed at Hewlett Packard and a version of it is applied by Motorola and available as a design tool in their HDC family of High Density CMOS gate-arrays. As mentioned above, the double-strobe flip-flops can be applied in Boundary-Scan cell designs to allow path-delay measurements next to the normal PCB interconnect tests.

Chapter 4

BST DESIGN LANGUAGES

This chapter describes various software tools developed to support automation in Boundary-Scan testing. These tools support the features of Boundary-Scan applications in various phases of the product life cycle, where 'product' may be defined as IC, PCB or system. Table 4-1 intends to place the various tools in their right context, which may be useful when reading this chapter.

Table 4-1 General and BST Software Support Tools

	Documentation/ Description	Design	Design Verification	Test Preparation
IC	BSDL (VHDL)	Automatic BSC inserter (TIM / BSR compiler)	Conform-ance with Logic Analyzer	ATPG for inserted BS logic
PCB	EDIF	--	Disassembly	BS Test Pattern Gen.
System	--	--	Disassembly	BS Test Pattern Gen.

BSDL DESCRIPTION

The Boundary-Scan Description Language (BSDL) is formally presented as Supplement (B) to the IEEE Std 1149.1-1990, [18] and [19].

BSDL describes the testability features of Boundary-Scan devices which are compatible with the IEEE Std 1149.1. It is written within a subset of the VHSIC Hardware Description Language VHDL [20]. As such, BSDL is in itself not a general purpose hardware description language, but it can be used in conjunction with software tools for test generation, analysis and failure diagnosis. The parameters used in a BSDL description are orthogonal to the rules of the IEEE Std 1149.1. This means that elements of a design which are absolutely mandatory for the IEEE Std 1149.1 are *not* included in the language. This avoids ambiguous

85

descriptions and definitions. For example the BYPASS register is fully and without options described in the IEEE Std 1149.1 and hence it is not described in BSDL. Also, BSDL is *not* intended to describe (parts of) the on-chip system logic, but merely the properties of the boundary register with its terminal connections.

As stated above, BSDL comes as a subset of VHDL, in a case-insensitive free-form multi-line terminated form. The Backus-Naur Format (BNF) is used to describe the syntax. Comments are enclosed between a "--" and an EOL (end of line) character. For long strings a concatenation character "&" is used to break the strings in arbitrary but readable form. The concatenation character has no syntactical meaning and can be thought of as used in a lexicographical preprocessing step before the parsing process starts.

BSDL may be used in two environments: in a full or in a partial VHDL environment. In a full VHDL-based system the BSDL information is passed through the VHDL analyzer into a compiled design library, from where the boundary scan data are extracted by referencing the appropriate attributes. In the latter case only a limited set of VHDL syntax can be parsed. A fully integrated VHDL environment is supposed in the following descriptions.

BSDL comprises three main sections: the Entity, the Package and the Package Body. These "packages" remain constant along with the IEEE Std 1149.1. However, the user may add his own application specific "packages" to the standard formats.

The Entity Section

The Entity describes the Boundary-Scan parameters of a device's I/O ports and attributes in terms of VHDL, with the specific IEEE Std 1149.1 related definitions coming from a pre-written VHDL standard Package and Package Body. The Entity in BSDL has the following structure.

```
entity My_Ic is              -- an entity for my IC
    [generic parameter]
    [logical port description]
    [usage statement(s)]
    [package pin mapping]
    [scan port identification]
    [TAP description]
    [Boundary Register description]
end My_Ic;                   -- End description
```

This structure should be maintained with the order of elements as shown here. The elements will be addressed in the next subsections.

Generic Parameter

In BSDL the generic parameter is intended to define the various package options that a device may have. The parameter is used for a logical-to-physical relationship of the device's signals. It comes in the following form.

generic(PHYSICAL_PIN_MAP:*string:="undefined"*);

The parameter must have a name. The string is initialized to an arbitrary value ("undefined") to prevent any selection if the parameter is not bound to a value.

Logical Port Description

The device's system terminals are given meaningful symbolic names, which are used in subsequent descriptions. This allows the majority of the statements to be "terminal independent". Renumbering or reorganization of the terminals can be done without any repercussions. It also allows description of devices that may be packaged in various forms. Non-digital pins such as power or ground may be and should be included for completeness. The form is as follows.

```
port(<PinID>;<PinID>;...<PinID>);
<PinID> ::= <IdentifierList>:<Mode>  <PinType>
<IdentifierList> ::= <Identifier> | <IdentifierList>,<Identifier>
<Mode> ::= in | out | inout | buffer | linkage
<PinType> ::= <PinScaler> | <PinVector>
<PinScaler> ::= <Identifier>
<PinVector> ::= <Identifier>(<Range>)
<Range> ::= <number> to <number> | <number> downto <number>
```

The *<Mode>* identifies the system usage of a pin, with the following options:

in   : for a simple input pin,
out   : for an output pin which may participate in busses,
inout   : for a bidirectional signal pin,
buffer   : for an output pin that may not participate in busses,
linkage   : for other pins such as power or ground.

A *<PinVector>* is shorthand for a group of related signals, e.g. *Data(1 to 8)* denotes indexed data signals, from Data(1) to Data(8).

Use Statement(s)

The *use* statement identifies a VHDL package needed to define attributes, types, constants and others that will be referenced. The form is:

   use STD_1149_1_1990.all;   -- Get 1149.1 information

Because BSDL deals particularly with the IEEE Std 1149.1 Boundary-Scan testing of devices, this *use* statement is mandatory in BSDL.

Package Pin Mapping

In here the VHDL attribute and constant statements are used to list the package pin mapping. The format is (example values):

   attribute PIN_MAP of My_IC:entity is PHYSICAL_PIN_MAP;
   constant dw_package:PIN_MAP_STRING:=<MapString>;

The attribute PIN_MAP is a string which is set to the value of the parameter PHYSICAL_PIN_MAP, as named above. VHDL constants are then written, one for each packaging version that describes the mapping between the logical and physical device pins. Note that the type of the Constant must be PIN_MAP_STRING. An example of a mapping is

   "CLK:1, DATA:(6,7,8,9,15,14,13,12), CLEAR:10, " &
   "Q:(2,3,4,5,21,20,19,18), VCC:22, GND:11"

Note that the string is (arbitrarily) divided into two parts by the concatenation character "&", which has no further syntactical meaning. A BSDL parser will read the concatenated content of the string and match names like CLK with the names in the port definition. The symbol - number or alphanumeric identifier - to the right of the colon (:) is the physical pin (1) associated with that port signal (CLK). In a *<PinVector>* listing, like for *DATA*, a matching must exist between the *DATA* signal components and the list of pins between parentheses. In the example above the physical pin 15 mapped onto the signal DATA(5).

Scan Port Identification

The five attributes below define the scan port of the device (example values).

attribute TAP_SCAN_IN of TDI:signal is true;
attribute TAP_SCAN_OUT of TDO:signal is true;
attribute TAP_SCAN_MODE of TMS:signal is true;
attribute TAP_SCAN_RESET of TRST:signal is true;
attribute TAP_SCAN_CLOCK of TCK:signal is (17.5e6, BOTH);

Obviously, the names TDI, TDO, TMS, TRST, TCK and/or any others must have appeared in the port description. The Boolean assigned is arbitrary; the statement is used to bind the attribute to the signal. The real-number field in the TAP_SCAN_CLOCK attribute denotes the maximum operating frequency of the TCK. The second field may read LOW or BOTH, specifying the state(s) in which the TCK may be stopped without the loss of data in the boundary scan mode.

TAP Description

This subsection describes a major part of the BSDL Entity structure, that is the *device-dependent* characteristics of the TAP. It may have four or five control signals, a user defined instruction set (within the IEEE Std 1149.1 rules) and a number of data registers and options.

The TAP Instruction Register (see chapter 2) has a length of at least two bits. A designer may add optional instructions as defined by the IEEE Std 1149.1 and/or add design specific instructions. Unused instruction bit patterns must default to the BYPASS instruction. The standard allows private instructions which need not to be documented, except if they can create unsafe conditions, for example board level bus conflicts.

BSDL uses six attributes to define the Instruction Register. These are given below in a descriptive manner, together with an example.

1.   *attribute INSTRUCTION_LENGTH of My_IC : entity is <integer>;*

     Example:

     attribute INSTRUCTION_LENGTH of My_IC : entity is 4;

     The Instruction Length attribute defines the length that all opcode bit patterns must have.

2.   *attribute INSTRUCTION_OPCODE of My_IC : entity is <OpcodeTable>;*

     Example:

     attribute INSTRUCTION_OPCODE of My_IC : entity is

```
"Extest (0000)," &
"Bypass (1111)," &
"Sample (1100, 1010)," &
"Preload (1010)," &
"Hi_Z (0101)," &
"Secret (0001)";
```

The Instruction Opcode attribute is a BSDL string containing the opcode
identifiers and their associated bit patterns. The rightmost bit is closest to the
TDO. The Extest and Bypass bit patterns are mandatory within the IEEE Std
1149.1. The Sample instruction is also mandatory but the bit patterns are not
prescribed. Note that other bit patterns may also decode to these same
instructions.

3.   *attribute INSTRUCTION_CAPTURE of My_IC : entity is <Pattern>;*

Example:

    attribute INSTRUCTION_CAPTURE of My_IC : entity is "0101";

The Instruction Capture attribute string states the bit pattern that is jammed into
the shift register portion of the Instruction Register when the TAP controller
passes through the *Capture-IR* state. This bit pattern is shifted out whenever a
new instruction is shifted in and the IEEE Std 1149.1 mandates that the least 2
significant bits are "01". Note that this captured data can become a valid
instruction if the *Capture-IR* state is left through the *Exit1-IR* to the *Update-IR*
state. But when the TAP Controller enters eventually its *Test-Logic-Reset* state,
the BYPASS instruction code is forced into the latches at the parallel outputs of
the Instruction register, or the IDCODE instruction, if it exists (see chapter 2).

4.   *attribute INSTRUCTION_DISABLE of My_IC : entity is <OpcodeName>;*

Example:

    attribute INSTRUCTION_DISABLE of My_IC : entity is "Hi_Z";

The optional Instruction Disable attribute identifies an opcode that disables a
boundary scan device. The BYPASS register is then placed between TDI and
TDO. This attribute allows the opcode to be identified for software use.
Many devices have this capability because it is very useful in testing practices.

5.   *attribute INSTRUCTION_PRIVATE of My_IC : entity is <OpcodeList>;*

Example:

attribute INSTRUCTION_PRIVATE of My_IC : entity is "Secret";

The optional Instruction Private attribute identifies private opcodes which may be unsafe for access. Software can monitor the Instruction Register to issue warnings if a private instruction is being loaded in run time.

6.   *attribute INSTRUCTION_USAGE of My_IC : entity is <UsageString>;*

Example:

attribute INSTRUCTION_USAGE of My_IC : entity is
    "Runbist (registers Boundary, Signature;" &
        "shift Signature; result 0011010110000100;" &
        "clock TCK in Run_Test_Idle; length 4000 cycles)," &
    "Intest (clock SYSCLK shifted)," &
    "MyBist (registers Seed, Boundary, Bypass;" &
        "initialize Seed 00110101; shift Bypass; result 1;" &
        "clock SYSCLK in Run_Test_Idle; length 125.0e-3 seconds)";

The optional Instruction Usage attribute gives additional information about the operation of an instruction. The types of information needed are register-, result- and clock-information. Note that for the RUNBIST a second register (Signature) will be placed between TDI and TDO, next to the Boundary register. When the test is completed, the Result is shifted out from Signature and should match the given pattern, where the rightmost bit is nearest to the TDO. The test is run during 4000 TCK cycles with the TAP Controller in its *Run-Test-Idle* state. This procedure complies with the standard, see chapter 2: The RUNBIST Instruction. The Intest usage denotes a shifting of the SYSCLK signal. For MYBIST three registers are used, Seed, Boundary and Bypass. The Seed register needs an initialization pattern. When the test is done, the BYPASS register's content should be "1". The test run lasts 125 milliseconds while the TAP Controller is in its *Run-Test-Idle* state.

Boundary Register Description

This is probably the most determining part of the BSDL Entity structure to be described. The IEEE Std 1149.1 Boundary-Scan [1] describes in its chapter 10 as much as fifteen design examples of Boundary-Scan cells. The original BSDL description ([18] and [19]) refers to these figures in the Standard by their numbers instead of repeatedly redrawing them in the text. The notation used is the following:

f10-16[1]   : denotes *figure* 16 in chapter 10 of reference [1],
f10-19c[1]   : denotes the *control* part of the device in figure 19,
f10-22d[1]   : denotes the *data* part of the device in figure 22.

Out of all the possible options in chapter 10 of the IEEE Std 1149.1, BSDL needs only six functionally different cell options to completely describe a boundary scan register. For each option the related figure numbers are given here.

BC_1   : f10-12, f10-16, f10-18c, f10-18d, f10-21c, an universal cell that can be used at input and output pins and supports all instructions.
BC_2   : f10-8, f10-17, f10-19c, f10-19d, f10-22c, an input cell with parallel output latch.
BC_3   : f10-9, an input cell without parallel output latch.
BC_4   : f10-10, f10-11, an input cell allowing signal capture only.
BC_5   : f10-20c, a cell at a 3-state pin where output control comes from a system pin.
BC_6   : f10-22d, a combined input/output cell used at a bidirectional pin.

These notations are used also in the definitions of the Package and the Package Body. Figure 4-1 gives an implementation of a Boundary-Scan cell (see also figure 2-17) which can be applied at both input and output pins (cf. f10-12, f10-16).

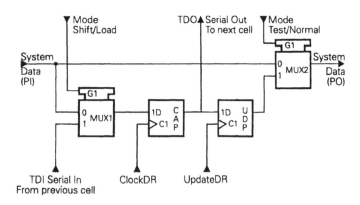

Fig. 4-1   An implementation of a Boundary-Scan Cell

The essential parts for description by BSDL are the system data input (PI) and output (PO), two multiplexers controlled by the instruction's Mode signal and the two flip-flops CAP and UPD. All other entities are precisely described by the IEEE Std 1149.1 and are omitted in the BSDL description. Given these considerations, the following three attributes suffice to describe the boundary register.

attribute BOUNDARY_CELLS of My_IC : entity is <CellList>;
attribute BOUNDARY_LENGTH of My_IC : entity is <integer>;
attribute BOUNDARY_REGISTER of My_IC : entity is <CellTable>;

Example:

    attribute BOUNDARY_CELLS of My_IC : entity is "BC_1, MyCell";
    attribute BOUNDARY_LENGTH of My_IC : entity is 3;
    attribute BOUNDARY_REGISTER of My_IC : entity is
        -- num (cell, port, function, safe [ccell, disval, rslt])
        "0 (BC_1, IN, input, X)," &
        "1 (BC_1, *, control, 0)," &
        "2 (MyCell, OUT, output3, X, 1, 0, Z)";

This example describes a 3-cell Boundary-Scan register. The first attribute shows a standard cell (BC_1) and a dedicated cell MyCell, which must have been described in a user defined Package. The second attribute defines the number of boundary register cells. This number must match the number of cells (string elements) found in the third attribute. The third attribute is a string of elements (one per cell), each consisting of two fields. The first field is the cell number, which must count from 0 to LENGTH-1. The numbering may be listed in any order. The second field consists of either 4 or 7 subfields between parentheses. The names of these subfields are listed in the comment line and their meaning is summarized below.

* The *cell* subfield identifies the cell used and must match a cell listed in the BOUNDARY_CELLS attribute.

* The *port* subfield identifies the actively driven or received port signal of the cell. An output control cell or an internal cell is indicated by an asterisk (*).

* The *function* subfield denotes the function of the cell, which can have one of the following values:

clock	:	a cell at clock input (f10-11)
input	:	a simple input pin receiver (f10-8)
output2	:	supplies data for a 2-state output (f10-16)
output3	:	supplies data for a 3-state output (f10-18d)
bidir	:	reversible cell for a bidirectional pin (f10-22d)
control	:	controls 3-state drive or cell direction (f10-18c)
controlr	:	a *control* that disables at *Test-Logic-Reset* (f10-21c)
internal	:	captures internal constants (cf. page 10-7)

Note that these are the functions of the boundary cell and *not* of the device's pin. An *internal* cell is used to capture constants (zeros and ones) within a design. Such cells must *not* be surrounded by system logic, cf. f10-7.

- The *safe* subfield defines the value to be loaded in the UPD flip-flop in cases where the software would otherwise provide a random value. The *safe* value may be used to prevent hazardous signal values being connected to system logic, for example during overdrive situations. An X signifies that it does not matter.

- The *ccell* subfield identifies the cell number of the cell that serves as an output enable.

- The *disval* subfield gives the value that the *ccell* must have to disable the output driver.

- The *rslt* subfield gives the state of a disabled driver, which can be either high impedance (Z) or a weak 1 (Weak1) or a weak 0 (Weak0). The latter two are used in ECL circuitry.

The last three subfields are needed when the cell function is either *output2* or *output3* or *bidir*. If the function is *bidir*, then disabling the driver means that the cell is a receiver.

The Package Description

As long as the IEEE Std 1149.1 remains unchanged, this BSDL Package can never be modified. However, users may create their own packages to define design-specific boundary cells in addition to the standard cells. In doing so, it is common practice to place the complete cell *descriptions* in the associated BSDL Package Body and only list the cell *names* in the Package.

The format is given here as an example.

```
package New_Cells is
     constant NC_1 : CELL_INFO;          -- New cell 1
     constant NC_2 : CELL_INFO;          -- New cell 2
end New_Cells;
```

Now *New_Cells* must appear as a Use Statement in an Entity description. Also all cell names must be named in the BOUNDARY_CELLS attribute string. The description of the deferred constants goes into the related Package Body.

The Package Body

As stated above along with figure 4-1, the BSDL description of a boundary cell comprises merely the parallel I/O, the flip-flops with their connections. Some further considerations to the Package Body can now be given with the aid of the symbolic model in figure 4-2.

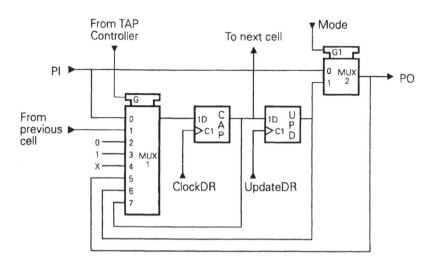

Fig. 4-2   Symbolic model of a boundary cell

Note that the standard allows for the omission of the UPD flip-flop in certain input cell configurations. Note also that the UPD flip-flops always get the CAP's data during the *Update-DR* state, so BSDL need not to describe this.

The CAP flip-flop may capture eight different signals from the previous multiplexer output. The multiplexer input signals '1' and '0' load constant values. The 'X' signal denotes don't-care. The latter may occur for example in an output cell during RUNBIST which loads a Linear Feedback Signature bit in the CAP flip-flop.

In VHDL, and hence in BSDL, a cell is defined in the Package Body as a *constant*. It is an array with the range unspecified, but the number of records is implicitly given in the *constant* definition. Each field of each record must be filled.

Example:

```
constant C_Ex_1 : CELL_INFO :=
 (  (Output2, Extest, One),  (Output3, Extest, One),
    (Output2, Sample, PI),  (Output3, Sample, PI),
    (Output2, Intest, PI),  (Output3, Intest, PI),
    (Control, Extest, One),  (Input, Extest, PI),
    (Control, Sample, PI),  (Input, Sample, PI),
    (Control, Intest, PI),  (Input, Intest, CAP) );
```

> The boundary cell design described here supports the EXTEST, INTEST and SAMPLE functions. The cell may be used as a simple function. It loads a '1' into the CAP during EXTEST if the cell is used as an output control function. During INTEST as an input, it reloads the CAP with the data value that was shifted into the cell.

It turns out that the symbolic model in figure 4-2 can be used to model all of the cells shown in the standard ([1]) except the one of f10-22d, because of its reversible nature. For that purpose the value *bidir* of the function as used in the above example for *BOUNDARY_CELLS of My_IC* is replaced by *bidir_in* and *bidir_out*, requiring a symbolic cell for each direction. The one to choose is defined by the value of the controlling cell. The values allowed for CAP data are PI, PO, UPD, CAP, ZERO, ONE and X. The values allowed for an instruction are EXTEST, INTEST, SAMPLE, RUNBIST, CLAMP and HIGHZ. Others such as BYPASS have no effect on the boundary cells.

Remnant BSDL Functions

This subsection describes a few remaining functions which are allowed options in the IEEE Std 1149.1.

ID Register Values

According to the IEEE Std 1149.1 the existence of either an IDCODE or an USERCODE (or both) implies that the optional IDCODE Register is available. Two attributes are needed to describe the instructions. Their format is given in the following example.

```
attribute IDCODE_REGISTER of My_IC : entity is
    "0011" &                -- 4 bit version
    "1111000011110000" &    -- 16 bit part number
    "00000000111" &         -- 11 bit manufacturer
    "1";                    -- mandatory LSB
```

attribute USERCODE_REGISTER of My_IC : entity is
"10XX" &
"001111001111 0000" &
"00000000111" &
"1";

The 32 bit partitioning of the IDCODE conforms to the standard, see also chapter 2. The rightmost bit is nearest to the TDO. The X value denotes a don't-care for the indicated bit position.

Register Access

The standard allows a designer to place additional data registers into a design. These registers are then accessed through design-specific TAP instructions. For the software it is important to know the existence of these registers and their length and associated instructions. Therefore this should be included in the BSDL description. The attribute for this is as follows.

attribute REGISTER_ACCESS of My_IC : entity is <RegisterString>;

Example:

attribute REGISTER_ACCESS of My_IC : entity is
"Boundary (Secret, User1)," &
"Bypass (Hi_Z, User2)," &
"MyReg[7] (LoadSeed, ReadTest)";

A seven-bit design-specific register *MyReg* has been added to the mandatory Bypass and Boundary-Scan registers. *Secret, User1, User2, Hi_Z, LoadSeed* and *ReadTest* must be previously defined user instructions.

The standard also allows user instructions to reference several concatenated registers at once, but then the concatenated register must have its own distinct name (see chapter 2). In the BSDL the concatenated register is treated as if it were a new register. The reason is that in any case, the data flowing out of a register is *not* known to BSDL, because BSDL is *not* a simulation language. Only the Boundary-Scan part of a design is covered by BSDL; the whole design is more properly the domain of VHDL itself.

Design Warnings

A designer may know circumstances where an IC Boundary-Scan design may cause hazardous situations when the IC operates in a board/system level environment. For example dynamic system logic may require sustained clock pulses when it is brought into its test mode, like INTEST. The Design_Warning attribute can be assigned a string message to alert future customers to potential problems. The attribute is given here as an example.

> *attribute DESIGN_WARNING of My_IC is : entity is*
> *"Dynamic device, maintain clocking for INTEST.";*

Note that the string is a textual message with no specified syntax and is not intended for software analysis. Moreover, the warning is tailored to specific display units only.

The next section gives an example how an ASIC vendor uses the BSDL in conjunction with the VHDL description of its system logic to *automatically* add the Boundary-Scan logic during the design phase.

BSDL DESIGN EXAMPLE

In the previous section it was pointed out that the BSDL can be used for *automated* design of a Boundary-Scan register around a designed core logic. Amongst others NCR/Teradata Joint Development has published a successful implementation [21] which will be summarized here as an example. The automated boundary register design has been applied to a number of ASIC CMOS gate arrays ranging in size from 4,000 to 80,000 gates with pin counts ranging from 160 to 364.

Design Methodology

The choice for a description in BSDL of the boundary cells was evident since the design of the core logic was already described in VHDL. Another consideration has been that BSDL supports the standard IEEE Std 1149.1 Boundary-Scan Test methodology, which has become common practice in many factories. And, in applying the IEEE Std 1149.1, a uniformity in the scan designs over the various ASICs is assured at the same time. Moreover a standard Boundary-Scan design needs to be developed once and is applied to all ASICs, which saves manpower.

The basic design approach is first to design the "naked" core logic as an hierarchical block, with inputs, outputs and internal functions, all described in VHDL. Then, in a second design/compilation step, the boundary cells are added to the design, described in BSDL. One condition is special in this example. An already present commercial logical synthesizer has been applied, which allows the reuse of intermediate files in the compilation process.

The next section shows a block diagram of a Boundary-Scan Register compiler. The program used here is called "Make_BS". The working method is as follows.

Make_BS assembles the required boundary cells from a library. It connects the cells between the core and the input/output pin buffers of the Boundary-Scan cells. It also connects the serial shift path from the TAP Controller, through all the Boundary-Scan Cells and back to the TAP Controller. It also connects signals such as Shift, Clock and Mode from the TAP Controller to the Boundary-Cell cells. Also provisions can be made for including ENABLE cells in the boundary chain to control groups of cells separately. This is done for loading and speed considerations.

At the start of the program, Make_BS generates a first-pass BSDL description along with the design of the boundary register. The initial ordering of the Boundary-Scan cells is arbitrary, but usually follows the VHDL port statements. Default Boundary-Scan cells are assigned, based on whether the pin is an input, output, bidirectional or a 3-state output. The design of the TAP Controller is standardized for all ASICs in the same family.

At this point use is made of the intermediate file from the commercial VHDL compiler. The designer may want to edit the first-pass BSDL file, which mostly consist of character strings. For example a different number ordering of the Boundary-Scan cells in the scan chain can be entered simply by changing the order numbers in the BSDL BOUNDARY_REGISTER attribute's first field (see above in previous section). At the same time, in the second field, the default types of boundary cell can be adapted to the wishes of the designer, e.g. BC_3 replaced by BC_1 along with other changes in this field's entries.

Then, Make_BS is re-executed and a new BSDL file is produced identical to the input file except that the BOUNDARY_REGISTER lines are numerically sorted to the number in the first field. A VHDL Package, containing component declarations to be used in simulation, is also produced.

The Circuitry

A simple octal D-latch is taken as a core to show how a Boundary-Scan chain can be added using BSDL and Make_BS. The following VHDL Entity defines the inputs and outputs of the core.

```
entity example_core is
    port(D: in bit_vector(8 downto 1); ENA: in bit; CLK: in bit;
        Q_OUT: out bit_vector(8 downto 1); Q_EN: out bit);
    attribute EXT_NAME of Q_OUT : signal is "Q";
    attribute ENABLE_LOAD of Q_EN : signal is 4;
end example_core;
```

The attribute EXT_NAME tells Make_BS the name to use on the outside of the shell if it is different from the core name. This prevents conflicts in naming conventions which may be imposed by a simulator. The attribute ENABLE_LOAD instructs Make_BS to use one boundary scan control cell for every 4 data cells which are controlled by Q_EN. This improves speed performance because of less output power needed per control cell. From this functional description the logic synthesizer makes a schematic diagram as in figure 4-3, together with the VHDL description.

Next the program Make_BS is started, using the obtained result of the octal latch. This results in a VHDL description of the core, the Boundary-Scan cells and the TAP Controller. A first-pass BSDL description is also produced.
The BOUNDARY_REGISTER part of the first-pass BSDL description is given below. The notation of the cell types (BC_1, BC_2 etc.) is the same as described in previous section.

attribute BOUNDARY_REGISTER of example : entity is

-- num (cell, port, function, safe [ccell, disval, rslt])

```
"019 (BC_3, D(8), input, X)," &
"018 (BC_3, D(7), input, X)," &
"017 (BC_3, D(6), input, X)," &
"016 (BC_3, D(5), input, X)," &
"015 (BC_3, D(4), input, X)," &
"014 (BC_3, D(3), input, X)," &
"013 (BC_3, D(2), input, X)," &
"012 (BC_3, D(1), input, X)," &
"011 (BC_3, CLK, input, X)," &
"010 (BC_3, ENA, input, X)," &
"009 (BC_1, Q(8), output3, X, 005, 1, Z)," &
"008 (BC_1, Q(7), output3, X, 005, 1, Z)," &
"007 (BC_1, Q(6), output3, X, 005, 1, Z)," &
"006 (BC_1, Q(5), output3, X, 005, 1, Z)," &
"005 (BC_1, *, control, 1)," &
"004 (BC_1, Q(4), output3, X, 000, 1, Z)," &
"003 (BC_1, Q(3), output3, X, 000, 1, Z)," &
"002 (BC_1, Q(2), output3, X, 000, 1, Z)," &
"001 (BC_1, Q(1), output3, X, 000, 1, Z)," &
"000 (BC_1, *, control, 1)";
```

At this point it is assumed that the designer wants to reassign the Boundary-Scan cell numbering, according to figure 4-4. The BSDL in edited in this way. Also the following changes are assumed to be necessary as well. For the clock input a "read-only" type of Boundary-Scan cell is required, referred to as BC_4. Also at the input pins, Boundary-Scan cells must be used with a parallel output latch, thus the default chosen BC_3 (without output latch) will be replaced by BC_1.

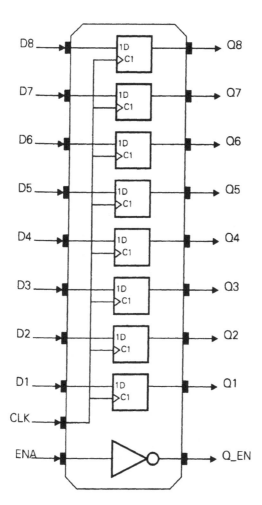

Fig. 4-3   Core of octal D-latch

The modified BSDL BOUNDARY_REGISTER part is shown in the next Attribute listing in which the changes are printed in **bold**.

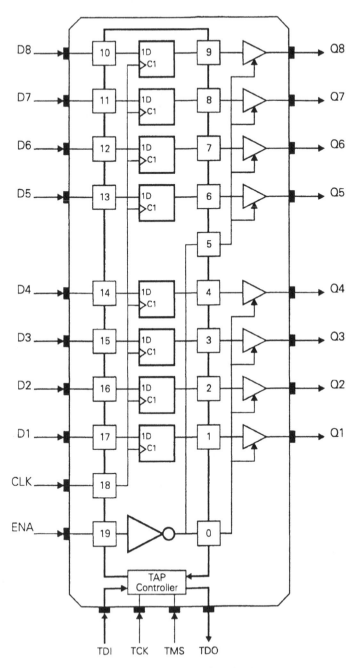

Fig. 4-4   Octal D-latch with Boundary-Scan chain added

attribute BOUNDARY_REGISTER of example : entity is

-- num (cell, port, function, safe [ccell, disval, rslt])

"010 (BC_1, D(8), input, X)," &
"011 (BC_1, D(7), input, X)," &
"012 (BC_1, D(6), input, X)," &
"013 (BC_1, D(5), input, X)," &
"014 (BC_1, D(4), input, X)," &
"015 (BC_1, D(3), input, X)," &
"016 (BC_1, D(2), input, X)," &
"017 (BC_1, D(1), input, X)," &
"018 (BC_4, CLK, input, X)," &
"019 (BC_3, ENA, input, X)," &
"009 (BC_1, Q(8), output3, X, 005, 1, Z)," &
"008 (BC_1, Q(7), output3, X, 005, 1, Z)," &
"007 (BC_1, Q(6), output3, X, 005, 1, Z)," &
"006 (BC_1, Q(5), output3, X, 005, 1, Z)," &
"005 (BC_1, *, control, 1)," &
"004 (BC_1, Q(4), output3, X, 000, 1, Z)," &
"003 (BC_1, Q(3), output3, X, 000, 1, Z)," &
"002 (BC_1, Q(2), output3, X, 000, 1, Z)," &
"001 (BC_1, Q(1), output3, X, 000, 1, Z)," &
"000 (BC_1, *, control, 1)";

This BSDL version is saved, compiled and Make_BS is run again. The result is a new scan register design and a new BSDL file. The new BSDL file is the same as the input file except that the BOUNDARY_REGISTER lines are numerically sorted in descending order of the (new) number in the first field (see figure 4-4).

A BOUNDARY-SCAN REGISTER COMPILER

This section describes a BSR compiler which has been developed by Philips Electronics [22]. The automatic generation of a Boundary-Scan Register around an ASIC also includes the generation of the test patterns for the boundary registers and the datasheet which complies with the IEEE Std 1149.1 for documentation requirements. The product is available from Philips under the name TIM (Testability IMprover).

Before going into the details of the compiler itself, first some characteristics and presumptions are given of the design environment in which the compiler is assumed to operate and which have been considered when the compiler was in its design phase.

- ASICs have short turn-around times (of several weeks) from design to silicon.

- The ASIC production quantities are low (100-10,000 pieces) as compared to standard components.

- A workstation design environment is assumed, giving support in schematic capture, logic simulation, timing verification and fault simulation.

- All input and output files should be in ASCII format as far as possible, to allow easy file transport between various design environments (workstations).

- Libraries of pre-designed functions for the core design should be available, containing basic elements such as AND and OR gates, flip-flops, latches, boolean functions or multiplexers, but also scalable structures like RAMs and ROMs or even microprocessor cores.

- A library of pre-defined building blocks for each IEEE Std 1149.1 Boundary-Scan Cell should be available.

- Documentation of the Boundary-Scan chain, as required by the IEEE Std 1149.1, should be generated automatically.

- Logic synthesis tools (synthesizers) should be available.

- A tool for automatic Test Pattern Generation (ATPG) should be available.

Since the boundary registers must comply with the IEEE Std 1149.1, the ASIC designer can be provided with standard elements to incorporate into the Boundary-Scan Test shell around the core. So manpower is saved and the ASIC designer needs not to have detailed knowledge of the Boundary-Scan circuitry.

The Technology Library

A Boundary-Scan Register library contains all elements necessary to create a BSR circuit around a core design, such as a gate array ASIC. A description file is used for each chip technology, e.g. CMOS or ECL, as input for the compiler.

The definitions of the constituting elements for this compiler differ from the ones used in the BSDL. Because the elements themselves are fully described in the IEEE Std 1149.1 with many possible implementation examples, they are listed here together with only a short description of each block (see table 4-3).

Table 4-3   List of BSR Library Elements

Name	Description
bs0bypa0	: Bypass register, 1-bit shift, capturing fixed '0'.
bs0cela0	: Boundary-Scan cell for uni-directional input or output pin, able to drive and sense.
bs0celb0	: Boundary-Scan cell for bi-directional pin, able to drive and sense.
bs0celc0	: Boundary-Scan cell for clock pin, able to sense only.
bs0ctra0	: Boundary-Scan cell for enable range of output buffers, able to sense and drive.
bs0idca0	: Device Identification register cell, capturing fixed '0'.
bs0idcb0	: Device Identification register cell, capturing fixed '1'.
bs0irga0	: Instruction register, 3 bit, selecting Boundary-Scan register (all modes) and Bypass register.
bs0irgb0	: Instruction register, 3 bit, selecting Boundary-Scan register (all modes), Bypass register and Device Identification register.
bs0muxa0	: Output multiplexer for TDO, used with bs0irga0.
bs0muxb0	: Output multiplexer for TDO, used with bs0irgb0.
bs0tapc0	: TAP Controller without reset pin.
bs0tapd0	: TAP Controller with reset pin.

As can be noticed, this description is slightly more detailed than the one used for BSDL, which defines only six building blocks (BC_1 through BC_6), but the similarity may be clear.

Both defined Instruction registers have a length of 3 bits. Table 4-4 lists the possible instruction values, the definition of the selected test data register and the test mode of the Boundary-Scan Test register.

Table 4-4   The Instructions and Their Effect

A B C	Selected test data register	Test mode of BS register
0 0 0	Boundary-Scan register	*EXTEST*
0 0 1	Device ID or Bypass register	*SAMPLE/PRELOAD*
0 1 0	Boundary-Scan register	*SAMPLE/PRELOAD*
0 1 1	Bypass register	*SAMPLE/PRELOAD*
1 0 0	Boundary-Scan register	*INTEST*
1 0 1	Bypass register	*SAMPLE/PRELOAD*
1 1 0	Bypass register	*SAMPLE/PRELOAD*
1 1 1	Bypass register	*SAMPLE/PRELOAD*

The rightmost bit is nearest to the TDO.

Compiler Input and Output Files

Figure 4-5 depicts the BSR compiler's input and output files, each of which will be described shortly. All files are in ASCII format to permit easy file transfer and to allow easy editing on every available text editor that can read in and output ASCII files.

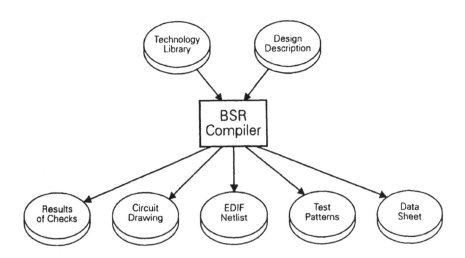

Fig. 4-5   Structure of BSR compiler

Design Description

The library of table 4-3 is used to build the Boundary-Scan register around the ASIC core design. Given the pin descriptions of the ASIC, the addition of Boundary-Scan cells can be done automatically because the cells are described in the IEEE Std 1149.1. The generated BSR information can subsequently be used to generate the test program for it and the standard prescribed documentation sheets. In the application described here, the design description input file consists of three parts.

1. The *header information*, which details items such as the date, designer's name, ASIC properties and name, whether or not a reset pin (TRST*) is needed for the TAP, whether or not a Device-ID register is present, etc.

2. The *pin description*, listing pin properties such as the pin number, pin name, pin type (needed for determination of a suitable Boundary-Scan Cell from table 4-3) and the chain position of the Boundary-Scan Cell. This list contains also the names of the control cells that enable the peripheral buffers for bi-directional and 3-state pins.

3. The *control cell description*, listing their properties.

Technology Library

Each technology for which a BSR library is available, is described in the technology description file. Two cell types are distinguished.

1. The *BSR cells* are specified in terms of name, search path, type, gate count, fan-in, fan-out, etc.

2. The *buffer cells* are specified in terms of name, search path, input load, output load, output drive, etc. The buffer description is used to select the appropriate buffer taking into account the number of cells driven and additional load due to wiring capacitance.

The 'Result of Checks' File

The BSR compiler starts with checking the consistency of the input files (design and technology descriptions) and any errors or warnings are logged in the 'result of checks' output file. If no errors are encountered, all other output files are generated.

The 'Circuit Drawings' File

This output file contains the commands for a schematic capture tool present for example on a workstation. Executing this output file results in a plotted drawing.

The 'EDIF Netlist' File

Besides the circuit drawing files, the BSR compiler also generates an alphanumeric netlist, containing the same interconnection information as used for the circuit drawings. The format supported complies with the EDIF 2.0.0 version [23]. A company proprietary netlist is also generated in this example.

The 'Test Patterns' File

The algorithm implemented in the BSR compiler's test pattern generator is, for all shift registers in the Boundary-Scan chain, based on the description of the 'flush test' [24]. Care is also taken that fault effects from the control logic of the shift registers (TAP Controller and Instruction register) are exercised and made observable through the TDO pin of the BST circuit by shifting out the same test data registers.

The 'Data Sheet' File

The data sheet output file contains all properties of the Boundary-Scan register and provides, if available, the values of the properties, all complying with the IEEE Std 1149.1 documentation standard. The structure and contents of this data sheet compares to the BSDL language description for Boundary-Scan devices.

Results

The test pattern generator makes no assumptions on the behaviour of the core. Therefore it is not possible to reach 100% coverage on the added circuitry. For example faults on data outputs of input cells, feeding into the core or faults on data inputs of output cells, driven by the core will not be detected. During other, functional tests these I/O connections are implicitly tested.

As an example an ASIC has been made; a server interface between a back-plane computer bus and a digital I/O system. The ASIC is a 1.2 micron CMOS gate array design with 5000 equivalent gates and 44 pins, including the TAP. The Boundary-Scan compiler has generated the BSR chain, the test patterns for it and a data sheet.

These test patterns are run on a fault simulator, simulating all single stuck-at input and stuck-at output faults of the BSR circuit. The results of the fault simulation are the following.

Number of circuit faults   :   2420
Number of inserted faults   :   2098
Number of collapsed faults   :   322

Hard detected faults   :   1849 = 76.4 %
Potentially detected faults   :   427 = 17.6 %
Undetected faults   :   144 =   6.0 %

Total fault coverage   :   94 %

Further analysis of the results showed that 340 potentially detected faults would be detected by the test equipment due to the applied 'flush test' algorithm applied within the test pattern generator. These faults concerned signals carrying the clock signal TCK. More than 100 of the 144 undetected faults were related to faults on the interface between the core and the Boundary-Scan circuitry. These faults are certainly detected in a full core test. Correcting the figures in the test results in order to meet these considerations, would lead to a fault coverage of 98%, which was considered to be sufficient.

TEST SPECIFICATION LANGUAGES

This section gives an overview of some widely spread specification languages, which all have in common that they are designed to support the test engineers and at the same time encouraging those engineers to apply Boundary-Scan Testing. Since in general parallel vector formats are easier to read for the test engineers, efforts will be spent to use the parallel format instead of the serial format. The next subsection describes such a parallel format, followed by a general format that is a merge of parallel and serial format. The last section describes a serial vector format which is commonly in use.

Test Interface Layer

The Test Interface Layer (TIL) is a test format capable of describing board tests, including BST, in a tester independent way. It is introduced by Philips Electronics. The languages described in the next two sections concern the BST Parallel Vector (BPV) lines which can include parallel vectors as well and the serial vectors (SVF) suited for BST. The language described in this section shows a higher level of abstraction: it describes a test procedure which also includes the PCB entities like nets, component pins and primary I/Os. The language is commercially available [25] and is called Boundary-scan Test Specification Language (BTSL). BTSL brings the description of serial access back to a higher level, in comparison with the parallel access as applied by the test engineer. In chapter 5 explain, along with the BST test flow, that the BST *test* process occurs as a parallel action, but that shift processes to enter the test stimuli and to extract the results are by nature a serial oriented process. From this point of view, the shifting process as such is not part of the actual testing activity on the PCB; shifting is only a *preparation* for test. At this level or application layer, here called the TIL [26], more capabilities of the test language are required including:

- test patterns -including for parallel access- relate to PCB entities,
- control of the test execution speed,
- permit loop constructs,
- allow timing dependencies.

The basic idea behind BTSL is that the printed circuit board properties need to be described only once, in a general data file, because it is the same for all tests, while each test set is described in a separate application data file. The information for the board description can be directly derived from the available netlists, component data sheets and/or other sources as needed. Both files are ASCII based and their structures are described below.

General Data File

This file comprises the following parameters.

- *File identification*
  This includes the BTSL syntax version number used in the file, the name of the PCB and its revision number. For example the general data file could start with:

  ```
  SYNTAX_VERSION 1.0
  DESIGN         EXAMPLE
  REVISION       1A
  ```

- *Board channel definitions*
  First the type of each channel has to be specified: SERIAL or PARALLEL. The serial type is of course BST based and entities to be stated are the number(s) and name(s) of each TAP with associated 'connection names' (TCK, TMS, TDO etc.), e.g. TCK3 is the test clock for TAP3, the TRST line (optional) for each TAP and finally the frequency is specified. This part of the general data file could read:

```
CHANNEL_DEF      SERIAL
  TAP1   =       X123
  TCK1   =       TCK_1;
  TMS1   =       TMS_1;
  TDO1   =       TDO_1;
  TDI1   =       TDI_1;
  FREQ1  =       0
END_DEF
```

A set of parallel channels is given an ID (also named TAP) followed by an equal sign ('=') and an integer stating the number of channels.

• *Serial (BST) channel definitions*
This part specifies the characteristics of the Boundary-Scan chain on the board, i.e. the order of the components in the chain, the available instructions for each component (defined including the instruction node), the selected scan registers and the default values for each register. A mandatory relation exists between a particular instruction and the associated data register. For example a BYPASS instruction should go to the bypass-register (BPREG). Background values can be specified for the bits in a data register. The compiler-tester inserts these values if no values have been defined in the application data file. For example the channel definition of the Boundary-Scan compatible octal buffer SN74BCT8244 of Texas Instruments (here called component U4) could read as follows (L=low or '0', H=high or '1' and N=don't care or unknown):

```
SERIAL_CHANNEL     TAP1
U4 NOTRST
   IR_LENGTH =     8            IR_CAPTURE   (HLLLLLLH)
   EXTEST          LLLLLLLL     BSREG  18    (NNNNNNNNNNNNNNNNNN)
   SAMPLE/PRELOAD  LLLLLLHL     BSREG  18    (NNNNNNNNNNNNNNNNNN)
   INTEST          LLLLLLHH     BSREG  18    (NNNNNNNNNNNNNNNNNN)
   TRIBYP          LLLLLHHL     BPREG   1    (L)
   SETBYP          LLLLLHHH     BPREG   1    (L)
   RUNT            LLLLHLLH     BPREG   1    (L)
   READBN          LLLLHLHL     BSREG  18    (NNNNNNNNNNNNNNNNNN)
   READBT          LLLLHLHH     BSREG  18    (NNNNNNNNNNNNNNNNNN)
   CELLTST         LLLLHHLL     BSREG  18    (NNNNNNNNNNNNNNNNNN)
   TOPHIP          LLLLHHLH     BPREG   1    (L)
   SCANCN          LLLLHHHL     BCREG   2    (NN)
   SCANCT          LLLLHHHH     BCREG   2    (NN)
   BYPASS          HHHHHHHH     BPREG   1    (L);
END_CHANNEL
```

• *Family definitions*
This is an optional section in the general data file. It specifies to which technology family a certain component belongs, e.g. TTL, ECL, CMOS, etc. This information can be used in the diagnostic part of the board tester to determine to which fault category a detected error belongs.

Application Data File

This file states the tests to be performed using the following parameters.

- *File identification*
  This includes the BTSL syntax version number used in the file, the name of the
  PCB, its revision number and the kind of test is specified. Optionally, it can be
  defined when a test set is to be marked as a FATAL_TEST if an error is to
  result in terminating the test for that board. For example the application data file
  could start with:

  ```
  SYNTAX_VERSION 1.0
  DESIGN         EXAMPLE
  REVISION       1A
  TEST           WALKING_ONE_TEST
  ```

- *Configuration Specification*
  In this section both the serial and the parallel test configurations are specified
  for the tester. More than one (partial) configuration may be required to test.
  Each configuration starts with a channel shift frequency (in kHz). The list of
  BST components is given for every channel (indicated by a TAP). For each
  component the required BST instruction is defined; the same applies to the
  default static data for the selected Boundary-Scan register. Optionally, a delay
  can be specified per channel for the update instant with respect to the update
  moment of the leading channel. For BST components that do not fully comply
  with the IEEE Std 1149.1, the leading and trailing TMS sequences can be
  specified which will be executed between the idle and capture states. These are
  known as the pause and update states, respectively. For parallel channels, the
  identifier (indicated by a TAP, e.g. TAP2) denotes a group of access nodes that
  have the same update timing. Again, a drive delay for this group can be
  optionally given. An example is given below:

  ```
  CONFIG              conf_1
     FREQ =           2000
     SERIAL_CHANNEL   TAP1
        TIMING =      0
        IC2  EXTEST   (LLLLLLLLLLLLLLLLLL)
        IC1  EXTEST   (LLLLLLLLLLLLLLLLLL)
     END_CHANNEL
     PARALLEL_CHANNEL      TAP2
        TIMING =      0
     END_CHANNEL
  END_CONFIG
  ```

- *Access Description*
  The access description lists the defined identifications for each configuration
  (access-id). Its use will become clear in the description of the Access Table.
  An entry for the access description may read as follows:

```
ACCESS_DESCRIPTION
   CONFIG  conf_1
      USE  A              ! 'A' is access_id
   END_CONFIG
END_DESCRIPTION
```

- *Access Table*
  The access table lists the board entities (nets, components, cluster inputs and control cells) that are used during the test. Each entry in the access table has an 'access_point' and one or more 'access_definitions'. The access_points are the mentioned board entities. The access_definition contains the identification (the access_id) and the prescriptions for the required parameter for both serial and parallel access. The serial access definition comprises a channel name (e.g. TAP1), the component's name (e.g. IC23) with (BS) cell number, the access type (R=read, W=write or B=both) and, optionally, the component's pin number. The parallel access identification includes the parallel channel identification (e.g. TAP2), the parallel pin number and an access type.

  For a sufficient test procedure, every bit of a (parallel) test vector (see test data specification) must match a board entity listed in the access table, in a one to one relationship. Next, in order to know which test vector should be applied to which board entity, the so called 'access-id' parameter is applied. The access_id is listed with each board entity of the access table and with each (parallel) vector. Using this access_id it is possible to match a test vector to a board entity or, in other words, to have a board entity tested only by those test vectors having the same access_id. Thus the Nth board entity in the access list is sequentially tested with the bits of the Nth column of the concerning test vector set.

  A part of an access table might look as follows:

```
ACCESS_TABLE
!net access_id channel comp   cell  type pin
!                             pin#
net1    A       TAP1    IC1    15    R    23
        A       TAP2           1     W;
net2    A       TAP1    IC1    14    R    22
        A       TAP2           2     W;
net3    A       TAP1    IC1    5     W    4
        A       TAP1    IC2    3     R    21;
END_TABLE
```

  Note that in this presentation every board entity (net) comprises two lines terminated by ';'.

- *Test Data Specification*
  This part of the Application Data File contains the (parallel) vectors to be applied to the specified configuration (USE_CONFIG). Each data line in this list consists of at least a vector number, the mapping reference (access_id, see above) and the actual vector. The vector options can include a LOOP construct,

a WAIT statement and a SIGNATURE definition. The value of the signature can be entered in the data file but also includes a LEARN option, to read the signature value from a known good device. Certain commonly used test patterns can be specified for vector sets (BLOCKs) by means of keywords such as WALKING_ONE, WALKING_ZERO, BINARY_COUNT, ALL_ONE, ALL_ZERO, RANDOM_SEED, etc. These keywords provide an automatically generated vector set instead of a vector set (determ_vector) which is entered by the user. Finally, a GUARDING option is available to every vector. Using this option, defined (static) values in the Configuration Specification (see above) can be overwritten. For an automatic generated test sequence of walking ones the test data part of the application data file may look as follows:

```
DATA
   USE_CONFIG  conf_1
001  USE A  WALKING_ONE
   END_CONFIG
END_DATA
```

Having seen all the above properties of BTSL, it can be stated that this test interface layer (TIL) resembles the description of in-circuit testing with the bed-of-nails fixtures. Moreover the description is tester independent.

General Format Test Vectors

With the application of BST as a method to test printed circuit boards (PCBs) the designer of PCBs was forced to pay attention to the overall design, validation and test strategy for the board at the same time. Furthermore a PCB seldom consists of full BST circuitry, testing such a board requires two types of test vectors: one for full BST components and one for conventional logic clusters.

On mixed boards the Boundary-Scan cells are thought of as virtual nodes, as opposed to the physical nodes which are accessed by the bed-of-nails fixtures. Testing such boards requires the generation of both the serial BST vectors and the parallel vectors for the bed-of-nails fixtures.

The output of a test pattern generator serving both serial (BST) vectors and normal parallel vectors becomes a mixture of these patterns. These mixed patterns can not be listed as one concatenated string. A more intelligent way of composing an eligible pattern has been invented by the Philips Electronics' Research Laboratories (CFT; Centre For manufacturing Technology). Due to the different nature of these two patterns, a new format was introduced, called the 'BST Integrated Testvector List' or BITL and has been used in the ESPRIT Project 2478: Research into Boundary-Scan Implementations. This vector list consists of lines called 'BST-Parallel-Vector' or BPV lines [27].

The BPV Line

The different nature of the serial BST test vectors and the 'normal' parallel vectors necessitates a new way of presentation. One of the main objectives in deriving a new form is the following:

If a general format can be found, then the test pattern generator will be tester independent, meaning that an already available tester can be used.

So nothing should hamper the introduction of BST in this case. Before describing the new format, one observation must be made first. A BST chain cycle consists of a *capture* action (sensing), followed by the necessary *shift* actions and is completed with an *update* action (driving). Considering these three actions separately, a tester can process patterns in a way that co-operates with the parallel vectors as follows.

1   shift in data AND drive 'virtual' pins

2   drive 'normal' pins

3   read 'normal' pins

4   read 'virtual' pins AND shift out data

In the new notation, these four actions and related data are represented as a single vector line: the *BST-Parallel-Vector* or BPV line. If more BPV lines are executed consecutively, each 'pair' of subsequent lines 4 and 1 combine to form a complete BST chain cycle, leaving only the very first and the very last BST lines of the total vector set of BPV lines 'single', i.e. meaningless.

```
1   shift in data AND drive 'virtual' pins
2   drive 'normal' pins
3   read 'normal' pins
4   read 'virtual' pins AND shift out data
1   shift in data AND drive 'virtual' pins
2   drive 'normal' pins
3   read 'normal' pins
4   read 'virtual' pins AND shift out data
. . . . .
. . . . .
. . . . .
1   shift in data AND drive 'virtual' pins
2   drive 'normal' pins
3   read 'normal' pins
4   read 'virtual' pins AND shift out data
```

In the time domain, the expected values can only be checked after the following vector, i.e. the following shift cycle, but the tester hardware or a post processor can take into account this time shift. The necessary timing aspects such as synchronous driving and sampling are not given here, but these remain the same as for a 'normal' vector.

Despite the presence of both serial and parallel test patterns, the BPV format is as usable as a 'normal' vector. However, extra control is needed for the generator of the BPV lines.

• The BST ICs have an Instruction Register used to place some ICs in the bypass mode whilst others are set in the self-test mode, etc. Therefore the pattern generator needs a configuration (CNF) line between the groups of BPV lines to control these instructions.

• When a particular state of the TAP Controller has to be entered, a TAP statement is needed to control the TAP signals (TMS and TDI) for every TCK. The inverted TAP (ITP) statement has been created for this purpose.

• Additionally, a reset statement (RST) has to be defined which resets the BST chain into a known state.

• The parallel bits in the parallel part of the BPV statement have still to be assigned to the nodes of the device under test (DUT). For this reason a parallel pin assignment line (PPA statement) is needed. The PPA statement assigns the values in the parallel field of the following series of BPV lines to the node names that exist on the DUT's netlist.

The description of the BPV lines is given in table 4-5 at the end of this section. Note that this table is meant as an example and does not state any specification.

Example

A simple example may clarify the application and the power of the BPV lines. The device under test consists of two ICs with BST capabilities. The circuitry has two input pins (PI1 and PI2), two output pins (PO1 and PO2), two nets (Net 1 and 2) that connect the ICs and a TAP, all as shown in figure 4-6.

It is assumed that the test plan of the example circuit prescribes that the following test patterns must be generated.

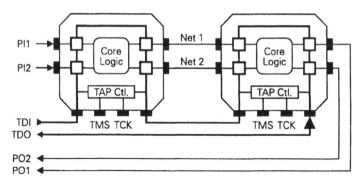

Fig. 4-6   Example circuit

• A reset instruction.

• An infrastructure integrity test (no ID registers). For this test an instruction register length of 8 bits is assumed for both ICs.

• A stuck-at test, consisting of all nets driving to '1', followed by driving all nets to '0'. This procedure makes the diagnosis simpler.

• A binary search for shorts.

Note that in the example circuit no control cells are included. If they were, their data would be placed in the BPV lines in the appropriate order and place. The following list gives the BPV test pattern list (see also table 4-5).

    ***   (Start)
    ***   (First reset all BST circuits and, at the same time, bring the TAP
           Controller in its *Test-Logic-Reset* state.)
    RST   **(T1, R1)**
    ***   Inform the tester which nodes are meant by the bits in the parallel fields
           of the following BPV vectors.)
    PPA   **(P, 4, PI1, PI2, PO1, PO2)**
    ***   (Now start the integrity test by shifting out the contents of the Instruction
           registers. According to the IEEE Std 1149.1 this starts with a '1''0'
           sequence ('1' is LSB). If an extra '1''0' is added to the contents, the
           connectivity of the TDI input is tested at the same time.)
    CNF   **(T1, 18, 01 XXXXXX01 XXXXXX01)**
    ***   (The actual board test starts here. Set ICs to *EXTEST* mode. (Note that
           the total Instruction register length is actually 16 bits.))

```
CNF  (T1, 16, 00000000 00000000)
***  (Set all nets to a '1' and test for stuck-at '0'.)
BPV  (T1, 8, 1111 1111; P, 4, 11 11)
***  (Test all nets for stuck-at '1'.)
BPV  (T1, 8, 0000 0000; P, 4, 00 00)
***  (Test for shorts. Note the shift order.)
BPV  (T1, 8, 0010 0111; P, 4, 00 11)
BPV  (T1, 8, 0101 1010; P, 4, 01 01)
BPV  (T1, 8, 1001 1001; P, 4, 10 10)
***  (End)
```

In this example the values for driving or sensing a node are mixed. To define whether a node must be sensed or driven, a second bit is needed adjacent to every bit in the pattern of the P-field in the BPV statement. Finally, it is necessary to have information about certain bits being a don't care or not, thus an additional third bit is needed on all nodes. The next example is a rewritten list for the complete test set, in which the comments are reduced and the two CNF statements are combined into one.

```
***    (Start)
RST    (T1, R1)
PPA    (P, 4, PI1, PI2, PO1, PO2)
CNF    (T1, 18, 460000004600000047)
***    (Stuck-at)
BPV    (T1, 8, 77337733; P, 4, 3355)
BPV    (T1, 8, 55225522; P, 4, 2244)
***    (Shorts)
BPV    (T1, 8, 55625766; P, 4, 2255)
BPV    (T1, 8, 57237532; P, 4, 2345)
BPV    (T1, 8, 75237523; P, 4, 3254)
***    (End)
```

The bit value strings given here are octal triplets (see next table).

Table 4-5 Description of the BTSL Instructions

***   (<comment>)
Comment line. The argument is:
comment = string of ASCII characters except the ')' symbol.

WTN(<ms>)
Wait for a given time before the next command. The argument is:
ms = time in milliseconds, $0 < ms < 65536$

(continued)

Table 4-5 (cont'd)   Description of the BTSL Instructions

CMD   (<arguments>)

CMD is a three letter command.

The following points hold for all commands.

- All I/O commands are assumed to start in IDLE state, except for @RST, @TAP and @ITP instructions following other @TAP or @ITP instructions.
- All I/O commands must end in this IDLE state except for the @RST command with reset mode 0 and @TAP and @ITP instructions that are followed by other @TAP or @ITP instructions.
- If a string is in reality shorter than a stated length, then the string will be extended to the correct length by repeating the last character.

RST   (T<t>, R<m> [;...])

Reset statement. The arguments are:

t = TAP number

m = reset mode in which

m=0 : soft rest (stay in *Test-Logic-Reset* state)

m=1 : soft rest (5x TMS HIGH on rising edge of TCK)

m=2 : hard reset (TRST LOW on rising edge of TCK)

m=3 : execute combination of m=2 and m=1

CNF   (T<t>, <l>, <s> [;...])

Configuration data for all BST ICs. The arguments are:

t = TAP number

l = length of instruction chain, 2 <l <65536

s = string of l ASCII characters. The rightmost character is sent first. Each character represents an octal data triplet:

b0 : Input data (TDI data to program Instruction registers)

b1 : Expected TDO

b2 = 1/0 : Care/Ignore for expected TDO

BPV   (T<t>, <l>, <v> [;...] [;P, <n>, <s>])

Test vectors for Boundary-Scan and primary I/O.

The arguments are:

t = TAP number

l = length of Boundary-Scan chain, 2 <l <65536

v = string of x characters (x ≤ l) for serial fields

n = number of primary I/O pins, 0 < n < 65536

(continued)

Table 4-5 (cont'd)   Description of the BTSL Instructions

s = string of 1 ASCII characters. The rightmost character is sent
first. Each character represents an octal data triplet:
b0 : TDI for serial field
        For parallel field: TDI or
                                      TDO if b1=1
b1 : Expected TDO for serial field
        For parallel field: b1=1 for drive
                                      b1=0 for sense
b2 = 1/0 : Care/Ignore for expected TDO

TAP   **(T<t>, <n>, <s> [;...])**
Direct TAP control. Every character in the string represents a TCK
cycle (a rising plus a falling edge). The arguments are:
t = TAP number
n = number of TAP instructions (characters), n < 65536
s = string of ASCII characters (x ≤ n) with HEXADECIMAL
representation. The rightmost character of the string is sent
FIRST! Each character represents a data nibble:
b0 : TDI
b1 : Expected TDO belonging to the falling edge (negative
        slope) of TCK, corresponding with this character
b2 = 1/0 : Care/Ignore
b3 : TMS

ITP   **(T<t>, <n>, <s> [;...])**
INVERSE TAP statement. The arguments are:
t = TAP number
n = number of TAP instructions (characters), n < 65536
s = string of ASCII characters (x ≤ n) with HEXADECIMAL
representation. The rightmost character of the string is sent
LAST! Each character represents a data nibble:
b0 : TDI
b1 : Expected TDO belonging to the falling edge (negative
        slope) of TCK, corresponding with this character
b2 = 1/0 : Care/Ignore
b3 : TMS

The Serial Vector Format

This format to represent BST test patterns was introduced by Texas Instruments
[28]. The Serial Vector Format (SVF) aims to increase the reusability and
portability of Boundary-Scan testing, through raising the level of abstraction for the

Boundary-Scan test descriptions. It is a kind of neutral format that can be used as a front end in various IEEE Std 1149.1 test applications. SVF also makes the test description tester independent.

SVF describes the state transitions of the IEEE Std 1149.1 bus in terms of transactions conducted between stable states. For example the scanning in of an instruction is described merely in terms of the data involved and the desired final state which the scan controller must eventually reach. Controller states like update, capture, pause etc. are inferred rather than explicitly described in a flat representation of tests.

SVF also supports cluster testing in a Boundary-Scan environment. As is known by cluster measurements through BST (see next chapter), the scan sequence necessarily includes a huge overhead of 'quiescent data' besides the wanted test data. The quiescent data are merely a filler of some digital value for the unused Boundary-Scan cells, in order to match the test pattern to the length of the Boundary-Scan chain. With SVF it is not necessary to specify every state of the IEEE Std 1149.1 TAP Controller at the lowest level to target one out of N Boundary-Scan cells.

The SVF statements come in an ASCII file format. For the complete specification is referred to [28].

Example

The test begins by disabling the TRST line and continues by setting the end states for both Instruction and Data scan cycles.
Next the headers and trailers for both the IR and DR scans are defined. The header pattern specifies how to pad the scan statement with a set of leading bits to accommodate the devices that are located in the Boundary-Scan chain *before* the component of interest. Similarly the trailer pattern specifies how to pad the scan statement with a set of trailing bits to accommodate the devices located on the scan path *after* the component of interest.

```
! Begin test program
TRST OFF;          ! Disable TRST
ENDIR IDLE;        ! End IR scans in IDLE
ENDDR DRPAUSE;        ! End DR scans in DRPAUSE
HIR          24    TDI(FFFFFF);  ! 24 bits IR header
TIR          16    TDI(FFFF);      ! 16 bits IR trailer
HDR          3 TDI(7) TDO(7) MASK(0);   ! 3 bit DR header
TDR          2 TDI(3); ! 2 bits DR trailer
SIR          8 TDI(41);    ! 8 bits IR scan
SDR          32    TDI(ABCD1234) TDO(11112222);          ! 32 bits DR scan
RUNTEST    95  TCK ENDSTATE IRPAUSE;      ! RUNBIST for 95 TCK clocks
```

```
SIR   8  TDI(00) TDO(21);      ! 8 bits IR scan and check value for status bit
STATE RESET;                   ! Enter Test-Logic-Reset
! End of test program
```

Given this imaginary example, a self-test of the concerned component is invoked by shifting in a value of hex 41 into the instruction register, followed by a hex ABCD1234 into the selected data register. Then the Run-Test/Idle state is entered for 95 TCK cycles after which the TAP controller is brought to the IRPAUSE (Pause-IR) state.

Finally a new instruction is scanned in and the status bits are compared with the expected values for further diagnosis.

Chapter 5

PCB TEST STRATEGY BACKGROUNDS

This chapter describes the basics behind the PCB test procedures. First the infrastructure of BST net itself is considered, which should be tested on its integrity before any other test is performed. Next the possible faults during the PCB manufacturing phase are described, followed by a discussion of various test pattern sequences and diagnostic algorithms required to perform board interconnect tests. Cluster testing is treated as a separate item in a following section. By searching the minimum test vector set for memory interconnect tests, it is proven that BST is an excellent tool for these type of tests as well. In the last section an architecture of a Boundary-Scan test flow is given.

TESTING THE INTEGRITY OF THE BST CHAIN

In electronic testing it is common practice to start every test procedure with the test of the infrastructure of the test circuitry itself. This holds also for the BST path. This principle should be followed in particular for testing PCBs on manufacturing defects, meaning that before using the BST capabilities for a PCB the BST access and interconnecting structures (net) should be tested first. This implies that all the whole interconnecting path between two 'virtual' test pins (BS cells in figure 1-23) must be considered as part of the PCB's test structure. It is assumed that the components mounted on the board are checked in advance, be it guaranteed by the supplier's outgoing test or as part of the incoming test before the PCB assembly.

In order to test the BST infrastructure and test capabilities, a test sequence is introduced [29], in which each test step relies on the results of the previous step(s).

The Tested Functions

Before describing the actual test sequence the faults to be covered should be defined first. For the infrastructure and the testability of the BST path the following functions need to be tested.

- The power supply lines to the components must be correct.

- The IC's BST implementations must be correct.

- The TCK and TMS signals must be present and be correct.

- The TDI and TDO signals must be present and be correct.

- If present, the optional TRST* signal must be correct.

- The TDI-TDO connections between the BST components on the board.

The effects of faults in the above points are discussed below. For the description of each point it is assumed that at any time only one fault is occurring. This single fault model comprises stuck-at and bridging faults. The open fault is considered as a stuck-at fault because of the presence of the pull-up resistors as prescribed by the standard.

Power Supply

A problem with the power supply itself is hard to check with BST and dedicated test equipment must be used to solve the problem. As to the power supply *lines* to the components a fault can result in a wrong voltage or no voltage at all at the component. If a BST component is not working as a result of a wrong supply voltage then the detection is straight forward. If a faulty voltage level causes an oscillating circuit on the PCB, then the results are easily detected as well, due to the unexpected measured responses. The diagnosis and location of the actual fault requires additional equipment and measurements.

BST Implementation in ICs

As stated above, it is assumed that only correctly functioning ICs are mounted on the PCB. However, during assembly of the PCB the IC is subjected to some hazards, for example from mechanical or heat treatments. But, while testing the infrastructure and the testability of a board, many of the IC's BST functions are used and tested implicitly, without any extra effort.

If the IC's BST logic is not responding properly, the TDO signal is either a '1', a '0', tri-stated or oscillating. The first two states are detected and localized normally. The tri-state is transformed to a logical '1' by the prescribed pull-up resistor at the next TDI input and is thus also localized normally.

An oscillating IC output results in an unexpected signal train and is detected with a high coverage. For example, suppose that 8 output bits are expected and that the oscillating process generates random values. Then the chance that such a value matches the expected value is $1:2^8 = 1:256$, in other words the coverage is more than 99.5%, which is assumed sufficient. If more ICs are connected after the oscillating one, the chances for a matching pattern are even much less and the coverage is assumed to be 100%.

The TCK and TMS Signals

If the TCK is stuck-at '1' or '0', then no action takes place in the BST structure and detection is simple: nothing works. If the TCK is bridged to another signal which possesses different values in time, then the TCK signal may be forced to that signal level and be changed to a wrong value at a certain time or cycle; that is if the other signal is 'stronger'. This is detected because an expected state transition occurs at the wrong time or not at all. If the TCK signal is stronger, then the bridging fault may not be detected until a functional test.

A stuck-at fault on the TMS net will also cease operation of the BST structure. If the TMS signal is bridged to another 'strong' signal, then the detection is the same as above for the TCK, unexpected vectors are measured. Similarly, if the TMS signal is stronger, then the bridging fault may not be detected until maybe a functional test. A stuck-at '0' fault will most likely result in a SHIFT-DR or PAUSE-DR state, because the standard requires that for these stages TMS is held at '0'.

A special case occurs when the TCK and the TMS signals are bridged. This fault is likely to occur at PCB assembly time, because in many BST compatible ICs the TCK and TMS pins are situated next to each other. So if a soldering splodge connects two IC pins and is present at either one of these two pins, chances are 50% that TCK and TMS are short circuited (bridged). In this case both signals will show exactly the same signal train, most likely the strongest of the two. This is again detected as a non-functioning BST IC. The result can be predicted, given the states of the standard TAP controller. If the TMS signal steps through a '010101...' sequence, then the states become, according to the state diagram (see chapter 2), CAPTURE-DR, EXIT1-DR, PAUSE-DR, EXIT2-DR, SHIFT-DR, EXIT1-DR again, etc. Now the test data are shifted out at a four times lower rate than expected, filled up with tri-stated TDO cycles that read as a '1'.

The TDI and TDO Signals

Stuck-at faults on the TDI and TDO nets are straight forward to diagnose because detection can be done to the level of the TDO-TDI connection from which the problem arises. Diagnosing the bridging faults to other nets is dependent on the IC technology used (CMOS, TTL), because either the '1' or the '0' will be influenced by the wrong signal. If the signal on the other net is 'weak' and not changing, the bridging error may not be detected at all in the integrity test. In that case the detection will take place during an interconnect test, where every net receives both logical values at least once. During the SHIFT-DR phase all signals are stable, thus also on the TDO-TDI connecting net. It is in this phase that a signal from the net to which the bridging exists can be detected.

The first TDI connection as seen from the tester to the PCB is tested by shifting in an additional '01' string which should appear again in the TDO signal.

Like the TCK and TMS signal pins, the TDI and TDO pins may also be adjacent on an IC package and are susceptible to bridging. In the *special case* that more ICs of the same type are used in a design, there is a chance that, for these ICs, after a reset action the same initial instructions are loaded and shifted out of each instruction register (IR) towards the board TDO pin. Assuming that three of these ICs are connected in series and have a five bit IR, then the following string may be present:

01xxx01xxx01xxx01

(the right bit is the LSB and 'x' denotes don't care). The last '01' allows the check for the first TDI pin on the board. Now, if the supplied test values are the same, xxx01, a short between the TDI and TDO of any BST IC remains undetected. A possible algorithm to check such a short between the TDI and TDO pin of an IC could be to provide a string which is different from all instruction register contents. The best option (see [29]) is to shift in an additional sequence '...1111100000...'. The numbers of '1s' and 0s' are equal and are the same as the longest IR in the BST chain on the board. So for a longest register of L an additional string of 2L bits is needed and the total length of the test string becomes 2L+C bits long, where C is the total length of all instruction registers in the BST chain. The fact that both '1s' and '0s' are added is to provide for a '01' occurrence, making the string technology independent (TTL, CMOS). The proposed sequence is not the shortest but guarantees detection of all possible faults, is easy to generate and simplifies the diagnosis.

The TRST* Signal

The optional TRST* line does not need much attention because the IEEE 1149.1 standard prescribes a reset procedure without it (see chapter 2). If this optional line is stuck-at '0', all ICs are constantly reset and nothing works. If a stuck-at '1' cannot be overridden by a tester signal, it will be left undetected until a TRST*-only is tried from a known controller state, e.g. at system power-up. A bridging fault of the TRST* to any other net may lead to unexpected resets of the connected ICs. The best way to detect a TRST* bridging fault is applying a 'weak' signal from the tester to the PCB TAP.

The TDI-TDO Connections Between the BST Components

When a development group has to design various ASICs for a particular PCB, it is very likely that one and the same BST circuitry is applied repeatedly in a number of ASICs. This may result in the same pin numbers being assigned to the TAP. The only difference in the ASICs from a point of view of board testing then comes from the other pins, their associated BST and Control cells and the identification registers. Now, if two similar looking ASICs are inadvertently interchanged during the PCB assembly, the infrastructure would work, but the interconnect test will most likely fail because of a mismatch between the different output and input connections. It is clear that checking for the (optional) identification registers provides a very cheap first stage check. If the ASICs are very much alike indeed, the PCB may even pass an interconnect test. This is not a drawback for the infrastructure test but the problem is that the PCB is manufactured and processed further and only the final functional test will show the error.

Conclusion

The discussion of the defined faults of the BST infrastructure indicates that a high coverage can be obtained with an integrity test. This test should be seen as a first and single test step. Because of the many technological circumstances under which a fault may show up, the test results make it often difficult to diagnose the exact cause of a fault. Not mentioned here are bridging faults to non-BST clusters. But since all nets to a cluster will be toggled at least once in a test procedure, the detection of such faults can be obtained in the same way as described above.

The Test Steps

The aim of the integrity test is to have a fast and easy means to check the BST infrastructure and testability. It is *not* the intention to fully test the BST functionality inside the ICs or on the board, like the Boundary-Scan data registers or the bypass registers. Problems in these registers will be detected during dedicated tests.

The integrity test sequence consists of four steps. Each successive step relies on a successful completion of the previous step. The four steps are:

1. Reset,

2. Instruction Register Shift,

3. Identification Register Shift,

4. TRST* Connection Check.

Each step is detailed below.

Reset

This step is needed because not all ICs have an internal power-up reset facility. A possible external power-up reset provision can not be trusted because it is part of the PCB assembly in the factory. The same applies to the optional TRST* line. The test prescribed by the IEEE Std 1149.1 (see also chapter 2) should be performed: keeping TMS high at a logical '1' during at least five clock pulses. Although not to be trusted either (faults in the TMS and/or TCK lines as described above), this is the only possible way to test the reset feature of the BST ICs. If a TRST* line is provided, the highest coverage for reset tests is obtained by combining two tests: the TRST* line kept low at a logical '0' while resetting according to the 'standard' method.

After application of this first step it is assumed that no faults are detected. Then the next test step can be taken.

Instruction Register Shift

If this second step is successfully completed, then it can be concluded that the TCK, TMS, TDI and TDO nets are connected properly. It can also be checked that all instruction registers (IRs) loaded the initial Capture value. Upon the Reset action, this value should select the IR or the bypass register if no IDCODE is present. During shifting through the instruction registers, a new instruction can be loaded in preparation for the next test step. Given the proposed test sequence above, the next instruction should select the identification register.

Identification Register Shift

If the former step failed, this step will add less value to the test result. In particular it is not certain whether the IDCODE instruction has been loaded through the initial reset or through shifting them in as a test code. It is better therefore to ensure that the previous step shows the expected results before continuing the integrity test.

Shifting out all ID codes provides information on the position and type of the ICs in the BST chain of the PCB. According to the IEEE 1149.1 standard those ICs that have no IDCODE implemented will shift out a '0' coming from the bypass register. If the test was successful, all connected ID and bypass registers are working properly and can be used for the last integrity test step.

Note that the test results of this step detect most of the faults that appeared in the previous test step. There are also economic reasons for ensuring that the results of the previous step pass. The previous test checks for bridged TDI-TDO or TCK-TMS nets. Such errors could result in removal of solder splodges, a relatively simple and cheap solution. Alternatively it may require the replacement of a surface mounted 244 pin ASIC, a job of a different order of magnitude.

TRST* Connection Check

The TRST* line was already used in the first integrity test step (for maximum coverage) but the result does not make sure that the TRST* connection works properly, because the 'standard' reset procedure of TMS at '1' during five TCK pulses was also applied. So a separate TRST* connection test is needed. Here again, the former test steps should have shown good results.

For this test each TAP Controller is brought to the Pause-IR state. Assuming the former test left the TAP Controller in its Run-Test/Idle state, then the TMS signal must have the following values at the rising edges of the successive TCK pulses: *11010* (see state diagram, chapter 2). Once in the Pause-IR state a reset signal is applied to all ICs connected to the TRST* line. Then these ICs are either reset or not, leaving them with one of three possible registers selected.

• If reset: the device ID or the bypass register is selected, which were also verified in the former test step.

• If the TRST* connection fails: the Pause-IR state is still on, no reset occurred and the selected instruction register is still selected.

When the results are shifted out, the number of shift cycles must match the length of longest register in order to make diagnosis possible. Shifting out the results occurs as normal in the TDI-TDO path. The use of presetting the TAP Controllers into the Pause-IR state can be demonstrated with table 5-1, showing the state changes as a response to TMS signals for the cases that the IC is reset and not (see also state diagram).

As can be seen from table 5-1, the ICs which could be reset have their Shift-DR state one TMS value later than that the *not* reset ICs reached their Shift-IR state. Comparing the TRST* test results with the expected values directly points to the location of a problem.

Table 5-1   TAP Controller states

TMS	State with TRST* reset	State with no TRST* reset
-	Test-Logic-Reset	Pause-IR
0	Run-Test/Idle	Pause-IR
1	Select DR-Scan	Exit2-IR
0	Capture-DR	Shift-IR
0	Shift-DR	Shift-IR
.	.....	.....
.	.....	.....
0	Shift-DR	Shift-IR
1	Exit1-DR	Exit1-IR
1	Update-DR	Update-IR
0	Run-Test/Idle	Run-Test/Idle

The following arguments may clarify last statement.

- The lengths of the expected register strings are known, even when a fault is expected. From the IC's data sheets the captured contents that can be expected from the instruction registers are known. The same applies to the ID codes. The probability that a wrong output stream of bits (no TRST* connection) matches the correct value is very small.

- If the TRST* connection fails, then the contents of the instruction register is output, of which the first bit is a '1'. So if a BYPASS bit ('0') is expected, this is detected and diagnosed immediately.

- If the TRST* connection fails, then the contents of the instruction register is output, of which the first bit is a '1'. If the contents of the ID register is expected, then the first bit is also a one. But the probability that an erroneous bit stream matches in total the IDCODE is virtually zero.

So any error in the TRST* connection will be detected.

Conclusion

The integrity test sequence provides a good insight of the PCB's infrastructure of the BST chain. The four steps of the sequence are related in that each successive step uses the results of the previous step. As a first test this sequence can save much time in the factory because it is attuned to the assembly process of the PCBs. The test does not cover any further functional tests. Because most PCBs are not fully (100%) loaded with BST components, the common way of testing through bed-of-nails fixtures should not be discarded. In fact, when this tool is available, it could be used to complete the integrity test of the whole PCB.

An Example

Based on an existing PCB of real complexity the BST integrity test has been performed. The total number of TCK clock cycles was calculated from which the test time followed.
The BST part of the PCB showed the following properties.

· *The board contains 12 BST ASICs.*
· *All ASICs have ID registers of 32 bits.*
· *All ASICs have a TRST* pin.*
· *All instruction registers are 3 bits long.*

For each of the four steps in the integrity test sequence the number of TCK cycles is determined.

Step 1.   To start with the stepping to the Run-Test/Idle mode takes one cycle. The reset procedure according to the standard takes five cycles, thus in total 6 cycles are needed here.

Step 2.   The length of all, and thus the longest, instruction cycles is 3 bits. So the number of IR shift cycles will be $2L+C = 2 \cdot 3 + 12 \cdot 3 = 42$. Additionally the TAP controller handling from the Run-Test/Idle state through shifting up to the Run-Test/Idle again, takes another 8 TCK cycles, which adds up to a total of 50 cycles for this step.

Step 3.   All, and thus the longest, identification registers are 32 bits, which means that $12 \cdot 32 = 384$ TCK cycles are needed. Additional TAP controller handling takes here 7 cycles, which adds up to 391 TCK cycles for this step.

Step 4.   A TRST* reset takes no clock cycles, but it will occur sometime in a TCK cycle, so 1 cycle is counted here. To diagnose the result, the longest expected register string ought to be detected, which here is the ID register, so $12 \cdot 32 = 384$ TCK cycles are used. Handling the TAP controller takes in total 10 cycles so that in total 394 cycles are needed for this step.

Adding up all the TCK cycles needed in the four steps yields 841 TCK cycles. When the TCK frequency is 1 MHz, then the total integrity test time is 0.84 milliseconds. If the clock frequency were 10 MHz, which is possible, then this time reduces to 84 microseconds. This time minimizes the chance that some hazards are imposed upon the circuitry just by testing the integrity of the BST net on the PCB.

PCB PRODUCTION FAULTS

In PCB production the most frequent faults are stuck-ats, opens and shorts. A short can exist between signal nets mutually and between signal nets and ground or power lines. Other common faults on PCBs are a wrong component insertion or a wrongly mounted component.

Unlike in IC testing, a faulty PCB in the production stage is usually repaired. Therefore board testing requires not only the fault detection but also a diagnosis of the fault. For this reason a fault model for PCBs must also include a translation mechanism to present the diagnostic results in an understandable representation for human interpretation.

A fault is said to occur when the measured signal value of a device does not conform to the expected value. In order to be able to diagnose the fault, a certain knowledge of the device under test is required. For example an 'open' fault on a PCB may translate into a logical '1' if TTL logic was used and to a logical '0' if ECL logic was used. The logic's technology must thus be known in order to deduce that a measured logical value of '1' means an open. Subsequently, when an open or a short has to be repaired during production, then it is also advantageous to have an understanding of the production process. For instance a bridging fault can be caused by a soldering splodge between two adjacent IC pins or an interconnection between two wires; an open may be caused by a missing soldering joint, a broken signal line (copper track on PCB) or a missing component.

Apart from these so called static errors intermittent errors may also exist. These errors must be detected and diagnosed as well. For repairs during production one should know that these kind of errors may be caused by a bad soldering joint or result from the extreme temperature sensitivity of a component.

It is these kind of considerations that have to be taken into account when possible PCB production faults are discussed.

Interconnect Structures

In figure 5-1(a) a BST test path between two ICs is depicted. In fact this is the most simple type of net appearing on a PCB to be tested. Some other BST net configurations are given in figure 5-1(b-d).

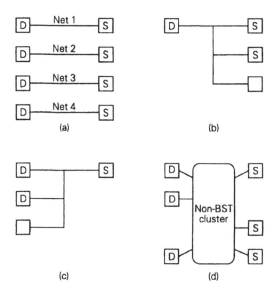

Fig. 5-1   Some PCB net structures

In this figure 'D' depicts a driving node and 'S' are sensing nodes; the unmarked nodes can refer to a bidirectional or a 3-state pin. Note that these nodes are 'virtual' nodes, that is they are supposed to be situated in the Boundary-Scan cells of ICs. Part (a) of figure 5-1 represents a number of simple nets, (b) is a multiple fan-out net, (c) is a multiple driver net (one driver at a time) and (d) is a net including a cluster. Because all indicated nodes are BST controlled, care must be taken that during interconnect testing that only one of the bidirectional or 3-state pins is driving the net at the same time and that bidirectional pins are included twice in a test: once as a driving node and once as a sensing node. The latter can be controlled and performed by means of the TAP controller. With respect to interconnect testing this means that for a BST net at least for each possible output node a test pattern should be generated. This will be clarified in a later example.

Finally, in figure 5-1(d) the cluster is a piece of non-BST logic. Functional testing of a cluster surrounded by BST logic is possible by considering the surrounding Boundary-Scan devices as virtual ATE channels. But it is obvious that net interconnect tests like in the other structures in figure 5-1 are not possible. The nets at the BST driving nodes D behave as if they were parts of the cluster's input circuitry.

The occurrences and detection of production faults are to a great extent technology dependent and diagnosis may become difficult. This is especially the case if multiple faults occur in connected nets where diagnosis may become even impossible. Moreover, it is not always clear which type of fault is responsible for the test pattern measured. Therefore some aspects of the various occurring fault types are considered separately in the following sections.

Considering Opens

The 'open' fault is frequently considered as one of the single-net faults which are categorized as follows.

- Stuck-at one   : the net is constantly at a logical '1' value.

- Stuck-at zero : the net is constantly at a logical '0' value.

- Stuck-open   : the net at an input pin is floating; the input may be updated to a '1' or a '0', dependent on the applied technology (e.g. TTL, CMOS). This fault is often simply called an 'open'.

Note that a stuck-at fault can influence the net as a whole, whereas a stuck-open fault may affect only part of the net. An input with a stuck-open fault may appear as an intermittent fault, e.g. due to induced hum voltage at a high impedance. In such a case the fault detection could be cumbersome and the fault model could be accused of malfunctioning. To avoid these possible occurrences, measures could be taken in the design phase (DFT); input buffers could be designed such that a predetermined value is applied when the input is left open.

The total number of stuck-open faults on a PCB not only depends on the number of (BST) nets but also on the number of nodes (IC pins) connected to the net. As indicated in a previous subsection, two interruptions on one net (PCB copper track) are not distinguishable without extra means. But in many cases there is also no use to it either because of the restricted area where such a fault appears. Usually a multiple open can be seen at one glance when one of the located interrupts is being repaired.

For the fault model a stuck-open fault is supposed to cause a partitioning of the net in two parts, a *subnet* as part of the original net and a leftover of nodes, forming the *rest*. The rest of the nodes may remain connected as a group or not. If a net comprises a number of p nodes (IC pins) then a partitioning of the net into two parts results in a subnet with either one node on its own or a node connected to a number of other nodes, to a maximum of $p-1$ nodes. So the subnet may count a number of i nodes varying from $i=1$ to $i=p-1$.

For each number i the number of possible combinations in a net with p nodes is given by:

$$C_i^p = \frac{p!}{(p-i)! \cdot i!} \quad .$$

To find all possible net partitions the combinations for all i from $i=1$ to $i=p-1$ have to be added. But now one has to realize that each combination appears twice because partitioning a group of distinct subjects into two subgroups of say a and b is the same as partitioning that group into subgroups b and a respectively.
Thus half the summation of all combinations of i has to be taken. This results in:

$$\frac{1}{2} \cdot \sum_{i=1}^{p-1} C_i^p = 2^{p-1} - 1.$$

The total number of partitioning a net of p nodes into two parts is thus given by $2^{p-1} - 1$. For example, if a net contains 4 nodes, then there are $2^3 - 1 = 7$ possibilities to partition the net into two parts; and a simple net between two nodes can only be split up into two parts.

Example

As an example a physical net as shown in figure 5-2 is considered.

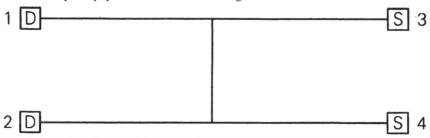

Fig. 5-2   Sample net with four nodes

The nodes 1 and 2 are indicated as D for 'driver' and nodes 3 and 4 as S for 'sensor'. As calculated above, there are seven ways to partition the net into two groups, a subnet and a leftover.

A number of these partitions is given in figure 5-3.

From figure 5-3 it can be seen indeed that a partition can be considered in two ways (which has led to counting only half the sum of all possible partitions). For example in case (a) in figure 5-3, the driving node 1 can be considered as the subnet and the other nodes as the leftover or rest of the original net. But if the part with nodes 2, 3 and 4 is considered as the remaining subnet of the original net, then the node 1 forms the leftover.

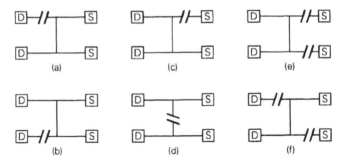

Fig. 5-3   Some single and multiple stuck-open faults

With respect to detecting and diagnosing the faults, from a closer study of all possible faulty connections it can be deduced that not all possibilities have to be taken into account. All stuck-open faults will be detected checking that the signal of each driving node can arrive at all sensing nodes. In practice this means that it is sufficient to check only those stuck-open faults which isolate a driving node (IC output pin) from the net, provided that all sensing nodes (IC input pins) on the net are being checked.

To come to the fault model it can be concluded that in case of opens it is sufficient when:

•   the number of tests (test patterns) equals the number of output pins, and

•   all receiving input pins on the net are read.

Considering Shorts

A bridging fault creates a short between two or more nets. Hence the name 'multiple-net' fault is sometimes used to denote a short, but usually a bridging fault is simply called a 'short'.

A short can exist between two signal nets mutually or between a signal net and ground or power supply. The short between ground and power supply is not considered here, because in power supplies no BST technology is applied. Other means are used to solve problems with the supply voltage.

The consequences of faults in the signal nets are a function of the applied technology. For example if one net with an output level of '1' is connected to a net with output at '0', then the resulting logical level at the shorted nets is not determined beforehand. If an output at '0' presents a lower impedance (is 'stronger') than a coexisting output at '1', then it can be expected that at a short between these two outputs the output signal of '0' prevails. In order to distinguish these situations, these types of short are usually given their own name.

- OR-short   : If the driving power on the net is such that a logical '1' dominates, then the resultant logic value is said to be an OR of the logical values of the individual nets.

- AND-short   : If the driving power on the net is such that a logical '0' dominates, then the resultant logic value is said to be an AND of the logical values of the individual nets.

- Weak short   : The resulting logical value on the net is not known but lies between '0' and '1'.

A connected driver or tester may impose the logic value upon a bridged net because of its very low output impedance. The resulting logic value on the net can also be undefined or non-deterministic. This may happen when the output impedances of the bridged outputs have about the same value, resulting in a weak short. Finally the test results are dependent on the various component technologies applied on the board. TTL IC outputs present a lower impedance at logical '0' output than at '1', while this is not true for CMOS ICs. But even when only one technology is used, it might be questionable what the final value at the net will be in case of a short. Suppose for instance that of an IC three adjacent TTL outputs are bridged of which two present a logical '1' and the other a '0'. The two '1'-outputs in parallel may provide a low enough output impedance to force the bridged net at '1', thereby overruling the lower impedance presented at the '0'-output. The latter could even burn out eventually because it can not drain the supplied current from the other two outputs.

Thought must therefore be given to the technologies used in order to be able to determine which shorts give what logic value at a certain net.

Shorts between signal nets and power or ground are considered separately. Given the extreme low impedances presented by ground and power, it is always true that a short to ground forces the whole signal net to zero volt and a short to power forces the net to the supply voltage. So, *independent* of the applied technology, if the one short (e.g. to power) results into a logic '1' then the other short (e.g. to ground) results in a logic '0'. In other words, the shorts to ground and power can be treated as stuck-at faults. The diagnosis of these faults can be made accordingly.

As stated above, at least two nets are needed for a short between nets, but shorts between more than two nets are possible of course. These are called multiple-shorts.

Figure 5-4 shows all possible shorts between four of the most simple nets, each one between two nodes (IC pins).

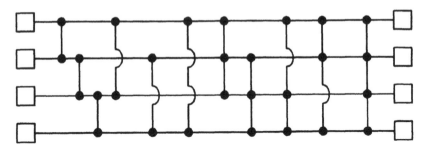

Fig. 5-4   All possible shorts between four simple nets

For two practical reasons it is considered acceptable if only the shorts between 2 nets are detected instead of all possible multiple-shorts.

1. A multiple-short may be considered as a superset of single 2-net shorts.

2. Testing for all possible shorts would require an exponentially growing number of test vectors.

This can be demonstrated by calculation as follows.
Consider a board with n nets, then the total possible number of i-net shorts is given by the combination of i out of n, which is:

$$C_i^n = \frac{n!}{(n-i)! \cdot i!} \cdot$$

For 2-net shorts this expression can be transformed to:

$$C_2^n = \frac{n!}{(n-2)! \cdot 2!} = \frac{1}{2}n(n-1).$$

To find all possible multiple-net shorts the combinations for all i from $i=2$ to $i=n$ have to be added. This leads to:

$$\sum_{i=2}^{n} C_i^n = 2^n - n - 1.$$

Here the exponential behaviour becomes already apparent. Table 5-2 shows the calculated number of shorts for some values of n. Table 5-2 shows clearly that the number of possible 2-net shorts remains reasonable whereas the number of total possible shorts rises drastically. Even at test frequencies of 10 MHz the figure of 10^{15} is extraordinary high to test them all.

Table 5-2   Comparison between 2-net and multi-net shorts

# of nets n	# 2-net shorts $\frac{1}{2}n \cdot (n-1)$	# total shorts $2^n - n - 1$
4	6	11
5	10	26
10	45	1013
15	105	32752
20	190	$\approx 1.05 \cdot 10^6$
50	1225	$\approx 1.13 \cdot 10^{15}$

Considering Multiple Faults

The BST technology offers full observability and accessability of all 'virtual' test nodes on the PCB. From this point of view, therefore, there seems to be no reason to restrict a fault model to single faults only, as is done in the IC technology for stuck-at faults. However, the amount of computer power and time to analyze all situations will rapidly grow to an unacceptable level. Even if this were reasonable, the question remains what the added value would be. In other words, is *substantial* information added to the single level fault model? And if not, is a partial multiple fault test the most economic solution? Is the diagnosis of multiple faults like stuck-at, shorts and opens still feasible? Can a *subset* of the *total possible* multiple faults on a PCB still offer a diagnosis with an acceptable fault coverage?

The solution for these questions comes from the knowledge of the characteristics of the physical layout of the PCB. It is important to know for instance that most production faults on a PCB appear within a restricted area. If a multiple short is localized by means of analysis of a two-short occurrence, then practice is that the soldering splodge connecting two signal nets also connects one or more other nets on the same spot. The multiple short will therefore be resolved at once at that spot. Likewise will a multiple open be located once a single open is localized; a scratch over the PCB surface has interrupted several adjacent copper tracks, some of which may belong to the same electrical net and others to a different net.

In conclusion it could be stated that the profit from extra computer power to diagnose multiple faults on a PCB should be weighed against the added value (fault coverage) to the results of the single fault model. Moreover, the present knowledge of the manufacturing process and the physical layout of the PCB easily disclose multiple faults once a single fault is localized. Multiple faults are discussed further in conjunction with confounding test results, in a later section. Some possible test results of combined faults on one net are considered in the section 'Diagnosing Interconnect Faults'.

Coming to a Model

Following the above considerations can form to a fault model that comprises the following points.

- Three fault types are to be detected and diagnosed:
  1. stuck-at faults: stuck-at 1 and stuck-at 0,
  2. stuck-open faults, preferably defined for the various technologies used in the ICs, such as TTL or CMOS,
  3. shorts, distinguished in or-shorts, and-shorts and weak shorts.

- Only those stuck-open faults are considered which isolate an output from a net.

- Only 2-short faults need to be detected.

- Multiple faults at one net are not considered, because at least one of the faults will be detected for sure. However, note that at PCB level the model should detect faults when they occur simultaneously on various, non interfering nets on a PCB.

This is only one possible scenario for a fault model and other options are equally valid. For instance, test schemes need not necessarily be attuned to a certain technology. Testing schemes exist that cover for example both AND-shorts and the OR-shorts, making the scheme more technology independent. One could also decide to extend the 2-short-only faults and design a test scheme that covers all possible multiple-shorts. The therefore extra computing power needed for the diagnostic capability can then be enhanced by incorporating the present knowledge of the applied technology in the ICs and on the PCBs. The ultimate choice of the fault model is usually a matter of factory management, meaning that design, production and marketing issues are considered before coming to a conclusion.

TEST ALGORITHMS

Knowing that the fault model allows diagnosis of the detected faults, the question now remains which test patterns need to be generated to cover all faults and yet make a diagnosis as simple as possible. Two observations will be made in advance.

- Although it seems obvious, an integrity test of the Boundary-Scan chain on a PCB should precede any Boundary-Scan testing procedure (see the first section in this chapter). The reason is clear, the on-board BST test facility is put on the PCB in the same production process in which the PCB is assembled. So only after testing the BS chain's integrity can the board interconnect test take place.

- For any separate net one and only one driving output must be active at any time. This is to avoid short conditions on the net. In particular this means that BST cells that control the enable signals of tri-state outputs or the signal direction at bidirectional pins, are never initialized such that two or more outputs are actively driving signals on the same net at the same time.

Especially the last observation can put a constraint on the BST procedure. To start with, care must be taken to load the right serial vector into the BST chain before the vector is applied to the net, which may take some extra time because the patterns have to be adapted to the PCB circuit structure. This does not influence the test pattern generation scheme because guarding is always needed and the total shifting times appear to be negligible.

The Binary Counting Test Sequence

It can be proven [30] that for a set of n nets $\lceil \log_2(n) \rceil$ vectors are necessary and sufficient to detect all possible shorts. The delimiters '$\lceil$' and '$\rceil$' denote the 'entier' function, meaning that the next higher whole number is to be taken if the worked out expression between the delimiters does not result in a whole number. This statement is demonstrated with the aid of figure 5-1(a), depicting a configuration of four simple nets. According to the above formula only two vectors are needed to detect any short. These vectors 'V' are composed as shown in the table 5-3. As can be seen from this table each net is assigned a unique binary number. Due to this assignment the binary value of the successive bits that form the test pattern differs per net. For example the test pattern fed to net 1 equals the binary value 0 and net 2 receives a value of 1. These patterns are fed to the driving nodes 'D' in figure 5-1(a). In case of a fault free set of nets, the same test patterns are measured at the sensing nodes 'S'. But in case of a short the sensed value differs from the driven value.

Table 5-3   Vectors for short detection

	V1	V2
Net 1	0	0
Net 2	0	1
Net 3	1	0
Net 4	1	1

The vectors applied in parallel to the four nets are called Parallel Test Vectors (PTVs) as opposed to the Sequential/Serial Test Vectors (STVs) applied to each single net over a period of time, through a number of PTVs. Note that the number of bits p of each STV equals the number of PTVs, hence the name: 'p-bit STV'.

The STVs and PTVs depicted in the above table are frequently called the 'vector set' which then is denoted as 'S' and presented as follows:

$$S = \begin{bmatrix} 0 & 0 \\ 0 & 1 \\ 1 & 0 \\ 1 & 1 \end{bmatrix}$$

If, in general, the vectors 'V' can be applied to all n nets in parallel, the test time is determined by $\lceil \log_2(n) \rceil$; if the test vectors are applied through a Boundary-Scan chain, then the test time is dependent on $p \cdot \lceil \log_2(n) \rceil$, where p is the number of shift positions.

To detect stuck-at faults a logical value is fed to the driving node which value should arrive at the sensing node. For example, in the fault free case, a driving pattern '01' should be detected as such at the sensing node. But if a '11' signal is detected instead, the associated net can be stuck-at 1 ('11' can also indicate an open fault). Similarly the reverse is true for a stuck-at 0. Therefore, if the test patterns from the last table were used to detect also the stuck-at faults, the test would fail for the nets 1 and 4. Clearly with the test pattern of net 1 a stuck-at 0 can not be detected and on net 4 a stuck-at 1 can not be detected. So if the all-0 and all-1 test patterns are avoided for the short test, then the same test vectors can be used to detect *both* the short *and* the stuck-at faults. As a consequence, instead of $\lceil \log_2(n) \rceil$ vectors then $\lceil \log_2(n+2) \rceil$ vectors are necessary and sufficient to test a set n of nets on both type of faults.

An algorithm to determine the required test patterns then reads as follows.

Step 1.   Assign each of the n interconnect nets a successive number, starting with 1.

Step 2.   Calculate the value of $\lceil \log_2(n+2) \rceil$ in order to find the number of patterns needed.

Step 3.   Assign each driving node the calculated number of bits with a bit pattern having a binary value equal to the number assigned to the net concerned.

The applied sequence of test patterns is also called the binary counting sequence, which is at the same time the minimum size needed for the detection of shorts and stuck-at faults. Of course, other test sequences exist as well, some of which are briefly mentioned later on.

A more elaborate example may illustrate the application of a binary counting or minimum-size sequence. Since this test pattern covers fully both the short and the stuck-at faults, it is said that the test pattern set has a fault coverage of 100% for these faults.

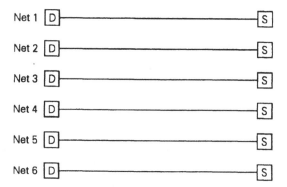

Net 1
Net 2
Net 3
Net 4
Net 5
Net 6

Fig. 5-5   A set of simple nets

The set of nets in figure 5-5 consists of the most simple nets, a connection between six driving pins (D) and sensing pins (S). Though very simple, the set of nets can be considered being general because it is stated that at any time only 1 driving pin is active. So the driving pins can in reality be a tri-state output or a bidirectional pin. Because 6 nets are involved, the required number of test vectors is

$$\lceil \log_2(6+2) \rceil = 3.$$

Table 5-4 shows the test patterns applied to each net which are building up a sequence of binary values corresponding to the net's decimal number.

Table 5-4   A binary counting test sequence

	V1	V2	V3
Net 1	0	0	1
Net 2	0	1	0
Net 3	0	1	1
Net 4	1	0	0
Net 5	1	0	1
Net 6	1	1	0

The test patterns applied to the nets exclude the all-0 and all-1 bit patterns. It can be seen that, for example, the driving pin of net 5 is driven with the binary value 5. Finally it can be seen that the applied binary counting sequence ensures that each net is toggled at least once between 1 and 0, covering at the same time also the test for stuck-open faults. In the practice of PCB testing this means that virtually all faults can be detected with the binary sequence test patterns.

The Minimal-Weight Sequence

This algorithm can be considered as a variation of the binary counting sequence. The designer determines the needed number of PTVs as described above in the binary counting sequence. Then the serial test vectors are generated such that first all STVs with only one bit with logical value '1' are produced, next those vectors possessing two '1s' etc. The value '1' represents the weight and since a minimum of ones are generated this is called a *minimal*-weight sequence.

Table 5-5 shows the idea for a set of 18 nets, for which, according to the previous section, four parallel test vectors are required.

Table 5-5 STVs generated in minimal-weight sequence

	4-bit STVs				Weight
Net 1	1	0	0	0	1
Net 2	0	1	0	0	1
Net 3	0	0	1	0	1
Net 4	0	0	0	1	1
Net 5	1	1	0	0	2
Net 6	1	0	1	0	2
Net 7	1	0	0	1	2
Net 8	0	1	1	0	2
Net 9	0	1	0	1	2
Net 10	0	0	1	1	2
Net 11	1	1	1	0	3
Net 12	1	1	0	1	3

The minimal-weight is used when no design and process information of the PCB is available.

The Walking One Sequence

Instead of the binary counting sequence the following bit stream can be applied to the Boundary-Scan chain for interconnect test purposes.

    ... 0 0 0 0 0 1
    ... 0 0 0 0 1 0

If there are N nets, then after N shifts of the total chain the logical '1' has walked over all nets, one at a time. That is why this test sequence is called the *walking one sequence*. Note that the total test sequence just takes N vectors (patterns), which is a parameter needed to determine the test time.

The walking sequence guarantees full diagnosis. For each test, the number of detected '1's is counted at the connected Boundary-Scan input cell(s). In the fault free case the number of received '1's should equal the expected number. Otherwise a net fault exists; a higher or a lower number of '1's is counted. A failing or unexpected '1' directly points to a faulty net.

A *walking zero* sequence (... 1 1 1 1 0) can also be used for the detection of faults.

In case of a walking 1 sequence, the total number of ones is increased with an OR-short and with a stuck-at 1 fault, whereas an AND-short and a stuck-at 0 fault reduce the counted number of ones. This indicates already that the walking sequences works perfectly for single-fault situations and for independent co-existing faults of the same type. But in the case of multiple faults of different types, two faults may mask each other. For example a stuck-at 1 fault raises the counted number of ones while a co-existing and-short decreases the same number by one. So the two faults mask each other.

The beauty of the walking sequence test method is that the test pattern is easy to generate and the test result is easily measured by a simple counter. So for a go/no-go test it is suited quite well, but the application time is still long. With respect to the pattern generation, the netlist information is still needed for the Test Pattern Generation.

The Diagonally Independent Sequence

If the set of test vectors in called S, then the general format of the diagonally independent test sequence is given by [33]:

$$S = \begin{bmatrix} 1 & x & x & x \\ 0 & 1 & x & x \\ 0 & 0 & 1 & x \\ 0 & 0 & 0 & 1 \end{bmatrix}$$

where x can be either a '0' or a '1'. The rows represent the STVs and the columns the PTVs. This implies that if b_{nm} is an element of $S(NxM)$ that:

$$b_{nm} = \begin{cases} 1 & \text{for all } n=m \\ 0 & \text{for all } n>m \\ x & \text{for all } n<m \end{cases}$$

An example of this sequence includes when case the for all x it holds that $x=0$, by which the walking '1' sequence is obtained.

This test sequence can be used for diagnosis of all type of faults, unrestricted shorts, opens, stuck-ats, etc. Moreover aliasing and confounding test results (see later on) can be eliminated with the aid of this sequence. Finally this sequence is systematically and easily to generate.

The Maximal Independent Set

This is a particularized version of the minimal-weight sequence. Apart from the weight (number of ones) of a serial test vector, a so called *potential weight* has been introduced [34], which depends on the bit positions of the highest and lowest bit with a value '1' in a serial test vector (STV). For a (non-zero) vector $v = (b_0, b_1, ..., b_n, ..., b_m, ...)$ where $b_n = b_m = 1$, the potential weight w is given by:

$$w = m - n + 1$$

For example, the STV '*0100100*' has a weight of 2 but a *potential* weight of 4. Vectors with only one bit ($m=n$) of value '1' have a potential weight of 1.

Table 5-6 STVs generated in minimal-weight sequence

	STVs (5-bits)					Potential Weight
Net 1	1	0	0	0	0	1
Net 2	0	1	0	0	0	1
Net 3	0	0	1	0	0	1
Net 4	0	0	0	1	0	1
Net 5	0	0	0	0	1	1
Net 6	1	1	0	0	0	2
Net 7	0	1	1	0	0	2
Net 8	0	0	1	1	0	2
Net 9	0	0	0	1	1	2
Net 10	1	0	1	0	0	3
Net 11	0	1	0	1	0	3
Net 12	0	0	1	0	1	3

After the determination of the number of required PTVs (see [34]) the generation of the max-independent vectors starts with the vectors possessing only one '1', then with those with two '1's, etc. Table 5-6 shows an example vector set for 12 nets. Note that the vector set exhibits a very regular pattern. The subsets with equal potential weight are diagonally independent.

The importance of this test pattern set is shown in a later section, when the diagnostic properties of vector sets are discussed.

The Order Independent Test Sequence

Yet another scheme of test patterns is the structure or order independent sequence, proposed in [31]. The idea behind this test sequence is the following. Consider all the input and output Boundary-Scan cells in a net like figure 5-5 on a PCB and suppose that this chain comprises N cells. To test the entire net with a test pattern according to the binary counting sequence (see above) in total a number of $\lceil \log_2(n+2) \rceil$ test vectors are needed, where n is the number of output cells. Assume a number of input cells is m and that $m+n=N$. Each of the test vectors is $n+2$ bits long and this vector is shifted into and applied through a chain of Boundary-Scan cells that is N cell long. Thus after the generation of the $n+2$ bits the test vector is padded with $N-(n+2)$ zeros to accommodate the total chain length of N bits. The zeros are padded in the proper order, *depending* on the order of the input and output cells in the Boundary-Scan chain. Thus the generated vector needs to be reformatted before it is shifted into the chain. So structural information is required about the configuration of the Boundary-Scan chain on the PCB.

This problem can be circumvented if the number of generated vectors is $\lceil \log_2(N+2) \rceil$ for a scan chain that is N cells long. N bits from the $N+2$ bits of each vector are loaded into the Boundary-Scan chain and the testing is performed as before.

The advantage of this system is that for vector generation no information about the PCB nets is needed and therefore this sequence is called the order or structure independent sequence. It is obvious that one disadvantage is the greater time complexity due to the longer vectors and therefore the sequence is not a minimum size sequence. Nevertheless, this vector set is more suitable for BIST implementation just because of its order-free test generation and loading property. However, the system only works if no 3-state pins are involved in the net. This makes the order independent sequence *not* suited for IEEE 1149.1 Boundary-Scan applications, because the 3-state control signals and the output values they control come through the same Boundary-Scan path. In order to prevent conflicts, the relationship between the control signal and the concerned output must be deterministic. Therefore the order independent sequence is not a realistic proposal for complex, bus oriented circuit boards.

DIAGNOSTICS

After each fault detection the test results have to be diagnosed in order to determine the cause of each of the detected faults. For the sample net of figure 5-5 the following number of different faults can occur:

- 6 stuck-at 0 faults
- 6 stuck-at 1 faults
- 6 stuck-open faults
- 15 2-net shorts (½· 6· (6−1))

The stuck-at and stuck-open types of faults are easy to pinpoint. But, in this simple example, a stuck-at fault may give the same diagnosis as the stuck-open fault if the stuck-open fault causes the open input to be forced to the same logical value as a stuck-at type of fault would cause (technology dependency). With respect to the 2-net shorts a complete analysis indicates that each of the possible shorts will give a unique test result, thus a unique diagnosis.

2-Net Shorts

Suppose for the nets in figure 5-5 that a short exists between the nets 1 and 2 and that the applied technology results in the type AND-short. Then applying the vectors to the net as indicated in table 5-4 will give the result as shown in table 5-7. The **bold underlined** printed digital values for vectors 2 and 3 are the **un**expected sensed values. The *combination* of these two faults is unique for the 2-net short between the nets 1 and 2. The determination of only one of the faults is not enough to conclude that a short between nets 1 and 2 exists.

Table 5-7 Test result for nets 1 and 2 shorted

Shorted nets 1+2	Driving signal	Sensed signal
Vector 1	000111	000111
Vector 2	011001	0**0**1001
Vector 3	101010	**0**01010

For demonstration the test results for a 2-net short between the nets 2 and 4 is given in table 5-8.

Comparing only the test results of vector 2 shows the same sensed signal values for both shorts, but in the latter case the combination with the test result of vector 1 makes the diagnosis unique again for the short between the nets 2 and 4. As with the stuck-at and stuck-open cases, the detection of all 2-net shorts may not be uniquely diagnosable because a stuck-at fault may cause the same test result. However, these faults may have been detected if the all-1s and all-0s test had preceded this 2-net short test.

Table 5-8 Test result for nets 2 and 4 shorted

Shorted nets 2+4	Driving signal	Sensed signal
Vector 1	000111	0000_11_
Vector 2	011001	0_0_1001
Vector 3	101010	101010

Note that thus far only 2-net and-shorts are considered. A closer study and analysis of multiple shorts will reveal that multiple-net sorts can mask 2-net shorts.

Multiple-Net Shorts

If, for example in figure 5-5 the nets 1, 3 and 5 were bridged, then the test result would be the same as for a 2-net bridging between the nets 3 and 5. The next table lists the result, in which the faulty bits are printed **bold**.

Table 5-9 Test result for nets (1), 3 and 5 shorted

Shorted nets 3+5 or 1+3+5	Driving signal	Same sensed signals	After repair of 3+5 short
Vector 1	000111	0001**0**1	000111
Vector 2	011001	01**0**001	01**0**001
Vector 3	101010	101010	101010

The multiple short 1-3-5 masks the test result of the 2-net short between the nets 3 and 5. However, if the latter is repaired, the short to net 1 then becomes apparent. This masking effect can be explained if one considers the sensed signals of vectors 1 and 2. In both cases the first *input* bit (which is applied to net 1) is a '0', leaving an AND-short undetected for diagnosis. The multiple-net short 1-3-5 also masks a possibly simultaneous stuck-at 1 fault of net 1 because, due to the short, net 1 will be pulled down to '0' as well. Here again, after removal of the short, the stuck-at 1 fault becomes apparent.

The masking of 2-net shorts by multiple-net shorts can be circumvented by using an additional set of vectors [32]. This set should consist of the complements of the former vectors, which are simply generated by inverting the binary values of the first vectors. Obviously, the required number of test vectors is now expanded to twice the above calculated number, that is to $2 \cdot \lceil \log_2(n+2) \rceil$, see table 5-10.

Table 5-10 The vectors of table 5-4 and their complements

	Original			Complement		
Net 1	0	0	1	1	1	0
Net 2	0	1	0	1	0	1
Net 3	0	1	1	1	0	0
Net 4	1	0	0	0	1	1
Net 5	1	0	1	0	1	0
Net 6	1	1	0	0	0	1

Using the same example once again, the complementary test vectors are applied to the set of nets in figure 5-5, assuming the same AND-shorts as before, i.e. a 2-net short between nets 3 and 5 and a multiple short between nets 1, 3 and 5. The test results are shown in table 5-11. It is clear that this (complementary) set of vectors uniquely identifies the multiple-net short. This will also hold in general, that is the two sets of test vectors give a full diagnosis of the shorts. Note that still only AND-shorts are considered.

Table 5-11 Test result for nets (1), 3 and 5 shorted

	Driving signal (compl.)	Results if nets 3+5 are shorted	Results if nets 1+3+5 are shorted
Vector 1	111000	110000	010000
Vector 2	100110	100100	000100
Vector 3	010101	010101	010101

From a symmetrical point of view the same reasoning is valid for OR-shorts. In the case of OR-shorts the complementary set of vectors would *not* distinguish the 2-net and multiple-net bridging faults as described above, but then this would be the case for the originally used test vectors. So the combination of the two complementary sets of vectors guarantees a full diagnosis for all possible shorts.

Still, not all possible PCB manufacturing faults related to shorts have been covered in the discussion so far. Two more aspects of bridging faults will be mentioned here: the aliasing and the confounding test results. A more elaborate treatment of these aspects is given in [33].

Aliasing Test Results

An aliasing test result exists when the faulty response of a set of shorted nets is the same as the fault-free response of another net. In this case it can not be determined whether or not the fault-free net is involved in the short.

Referring back to table 5-4 it can be seen that in case of an AND-short between the nets 3 and 5 the resulting serial test response of these nets will be *001*, which is also the test result of net 1 in the fault-free case. So, given the short between the nets 3 and 5, it can not be decided whether net 1 is also involved in the short or not. For net 1 the test result is the same in both cases.

Confounding Test Results

During manufacturing it may happen that two or more independent shorts occur in a set of nets on a PCB. A confounding test result may then occur when the test results from the multiple, independent faults are identical. In this case it can not be determined if these faults are independent.

Referring back to table 5-4 it can be seen that in the case of an AND-short between the nets 1 and 2 the resulting serial test vectors of these nets will be *000*, which is also the case if the nets 3 and 4 are bridged. Consequently it can not be determined whether there are two independent faults or that a single multiple-net short exists between the nets 1, 2, 3 and 4. A comparable result can occur if the above-mentioned additional complementary set of parallel test vectors is applied to the six nets.

Note that the example of a confounding problem is also encountered when only nets 1 and 2 have an AND-short and one of the other nets is stuck-at 0. Repairing the short would reveal the stuck-at fault with a subsequent test. The same conclusion could be drawn after seeing (during repair) that the 2-net short only existed between net 1 and 2.

The abovementioned AND-shorts between the nets 1+2 and 3+4 in the set of nets depicted in figure 5-5 are not the only occasions that the test result shows *000*, see table 5-4. In total six of those shorts give the same result: nets 1+2, 1+4, 1+6, 2+4, 2+5 and 3+4. Therefore it is said that the confounding degree C for this test result is six, C=6. Thus the degree of a confounding test result is defined as the maximum number of potentially independent faults that all have the same test result.

The above discussion should make clear that a full diagnosis after a one-step test procedure is only possible if neither a confounding nor an aliasing test result exists.

Single and Multiple Step Tests

Two types of test and diagnosis techniques are distinguished. The first one is the single-step test algorithm, where a set of test patterns is applied and the response is analyzed at once for fault detection and diagnosis. This is the type of diagnosis that is considered thus far in this section, because it is assumed that only a single fault in a set of nets will occur and that a unique diagnosis is possible.

In the multiple-step algorithm a test is applied, its response is analyzed and then one or more additional tests may have to be applied for a unique diagnosis. This procedure may be applied when aliasing (but not confounding) test results are obtained. Then a second test is needed to resolve whether or not the vector, to which it is aliased, has also failed. A parallel test vector (PTV) is applied such that the bits applied to the fault-free nets to which it is aliased, are set to '1'. The remaining bits are set to '0'. Clearly, if a net is part of an AND-short, the response bits are the same as the short and will be '0', otherwise they remain '1'. For example in the case of the aliasing nets discussed previously, the nets 3 and 5 were bridged and the test result of the AND-short was *001*, which aliased to net 1. In the second step of the two-step procedure an all-1 STV is applied to net 1 while nets 3 and 5 (amongst others) are supplied with an all-0 STV. In case of a 3-net short the response of net 1 will be *000*, in the fault-free case of net 1 it is *111*. This test can be performed for all distinct aliasing results in parallel, so that only one additional test step is required.

Confounding test results can not be resolved with this procedure, because that concerns two separate faults as opposed to a fault-free and a faulty net in the aliasing case. In order to fully diagnose also the confounding test results an adaptive test procedure is proposed in [33].

Adaptive Test Algorithm

Instead of diagnosing the set of all faulty nets, this algorithm analyzes the first test results, then the nature of the faults is determined and depending on this outcome it is decided whether or not additional tests are required. The following steps describe the adaptive test algorithm.

Step 1.   Apply the $\lceil \log_2(n+2) \rceil$ test vectors for fault detection.

Step 2.   Analyze the test results. If neither aliasing nor confounding results exist, then the diagnosis is complete and unique; the test is ready.

Step 3.   If only aliasing results are detected, then one more parallel test vector is required for full diagnosis, as described above.

Step 4.   If confounding results are detected, then also the presence of aliasing results must be considered. The confounding results are resolved as follows. Let C be the largest possible degree of confounding test results, as described above. For C confounding results no more than $C-1$ tests are required to resolve the confounding causing faults. Since the detection of these faults can be done in parallel, $C-1$ tests suffice to completely detect these faults, by which all the confounding problems are diagnosed.

Step 5.   As indicated in Step 4, after removal of the confounding causing faults, one or more aliasing results are still possible. As already stated in Step 3, one more test step is needed to finally resolve this problem.

In conclusion it can be stated that with this adaptive test algorithm an unrestricted full diagnosis of all shorts is possible and that no more parallel test vectors are needed than $C + \lceil \log_2(n+2) \rceil$, where C is the highest possible degree of confounding in a set of n nets.

Applying the Max-Independence Algorithm

When a highly loaded PCB contains many nets, say a couple of hundred, then the number C of the maximum possible confounding test results may become very high. Therefore, a unified test theory has been developed [34] in which the available technical knowledge of the PCB being tested is included to derive a full test and diagnosis. For instance it is assumed that shorts on a PCB are caused by soldering splodges and bridging occurs at *adjacent* nets. The theory proposes a more 'intelligent' test vector sequence: the maximal-independent sequence, described in a previous section, which then is applied to the nets, yielding the demanded full diagnosis. With respect to the binary counting sequence, this former sequence shows a slight increase of the number of required vectors. But the number remains far below the upper limit of n, which is required in the walking sequence test algorithm for testing n nets.

As an example assume a model in which at the utmost five adjacent nets out of say 200 nets are supposed to be bridged. The designer uses his manufacturing knowledge to determine that number of five, knowing that empirically 98% of the PCBs have never more than 5 adjacent shorts. This outcome determines the overall degree of diagnosis and if that is considered sufficient, the number p of test vectors can be determined (see below) following the assumption of maximal five adjacent shorts. The number of bridged nets is called the *extent* of the short, denoted as E. Note that the minimum number of $E=2$ and that the maximum number is $E=n$, where n is the number of nets. Now, it can be proven that, for the max-independence algorithm, the number of parallel test vectors for full diagnosis of the supposed shorts is given by:

$$p = \lceil E + log_2(n+2) - log_2E - 1 \rceil \qquad\qquad (a)$$

when the all-1 and all-0 sequences are excluded for reasons as before. For $E=2$ this number reduces to $p=\lceil log_2(n+2) \rceil$, the number for the minimum-size binary counting sequence for testing shorts. From the above it follows that for $2 \leq E \leq n$ the boundaries for the number p needed for full diagnosis are given by:

$$\lceil log_2(n+2) \rceil \leq p \leq n.$$

Within these margins the designer has to determine 'intelligently' the number of p parallel test vectors (PTVs) such that the overall diagnostic ambiguity of the test is minimized or, in other words, such that the diagnosis is considered sufficient.

The significance of the max-independence algorithm can be shown in an example. Suppose that on a PCB of 1000 nets a multiple-short of at maximum 20 adjacent nets covers 99% of the total number of all encountered shorts in the factory ($E=20$). According to the above equation (a) the number of PTVs then becomes:

$$p = \lceil 20 + log_2 1002 - log_2 20 - 1 \rceil = \lceil 20 + log_2 50.1 - 1 \rceil = 25 .$$

Thus the 25 PTVs fully diagnose these type of shorts in a population of 99% of all manufactured boards, which is considered sufficient. The upper bound is *1000* vectors, when the multiple-short over *all* the nets is to be considered, which is significantly more than the 25.

It should be remembered that prior technical knowledge is expected regarding the adjacent nets on the PCB as well as regarding the empirical maximum number of expected shorts occurring in the manufacturing environment. Note also that a net on a PCB can be physically adjacent to its neighbouring nets in more than one dimension, as is the case with multi-layer boards. Bare board testing, however, will eliminate most of the inter-layer shorts.

Combined Faults in One Net

Sometimes it may appear that the faults on a net can not be fully diagnosed with any one of the foregoing diagnostic procedures. The only way out then is to first repair one or more of the faults, followed by diagnosing the remaining fault causes. Some work has been done to obtain maximum diagnosis in these cases too [35]. In this section the problem is outlined using a few examples before coming to a conclusion.

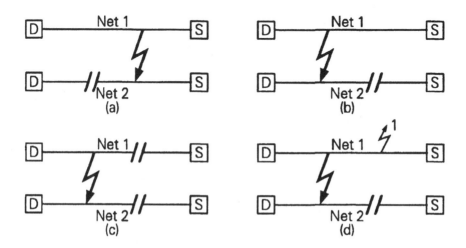

Fig. 5-6   Multiple faults in one net

The vector set applied to the two nets is the following:

$$\begin{bmatrix} 1 & 0 & 0 & 1 \\ 1 & 0 & 1 & 0 \end{bmatrix}$$

Four input PTVs p_n ($1 \le n \le 4$) are applied such that all possible binary values of the successive bit pairs are fed to the driving nodes 'D'. Vectors p_1 and p_2 apply the binary pairs '1' and '0' respectively to the driving nodes and p_3 and p_4 provide opposite bit values to the nodes. The parallel response vectors are denoted r_n. The serial response vectors are denoted s_1 and s_2 for Net 1 and 2 respectively. The responses for each of the examples (a), (b), (c) and (d) in figure 5-6 are discussed below. For the model it is supposed that the shorts are AND-shorts and that an open sensing node 'S' will be driven to '1'.

Case (a)   Both s_1 and s_2 exhibit '*1001*', because the sensing nodes are shorted and the driving signal at Net 2 is interrupted *before* the short, as seen from the driving node. Hence only the driving signal at Net 1 arrives at the sensing nodes. The open fault in Net 2 remains invisible. Only after removing the short, the open fault can be detected in a second test cycle.

Case (b)   The response vector set is now

$$\begin{bmatrix} 1 & 0 & 0 & 0 \\ 1 & 1 & 1 & 1 \end{bmatrix}$$

Clearly the response pattern s_2 shows an open, because irrespective of all input signals only the value '1' is sensed. In addition the short at Net 1 to Net 2 is also sensed because of the last '0' in s_1, which would be a '1' otherwise. Note that Net 2 is now interrupted *after* the short, as seen from the driving node. So for Net 2 it can be deduced that an open exists at the sensing node and that for Net 1 a short exists to Net 2. This means a full diagnosis.

Case (c)   As a matter of fact, *both* nets have a multiple-fault: a short and an open. The serial response vector at both sensing nodes is '*1111*'. At least this indicates that both the sensing nodes have an open fault, though a double stuck-at 1 could come to mind. Important to note here is that the short remains fully invisible, so full diagnosis is only possible after the opens are fixed. As a consequence two tests are needed.

Case (d)   The arrow to one depicts a stuck-at 1 fault, and is assumed a strong 'stuck-at' source. Here again each net possesses a double fault, a short plus a stuck-at and a open for nets 1 and 2 respectively. The serial response vector at both sensing nodes is '*1111*'. It will be clear that with one test cycle a diagnosis is impossible. First one or two faults must be solved before the remainder can be diagnosed. Anyhow, in every case more than one test is required.

Note that the pair of parallel test vectors p_3 and p_4 exhibit various test pattern sequences: a walking 1 sequence, a binary counting sequence, a diagonally independent sequence and a maximal independent sequence. From that point of view, when in the above cases a one-step test allowed a diagnosis of the short, this was at least due to the response of those two PTVs.

As a conclusion it can be stated that, with multiple-faults on one net, usually one or more faults must be solved before full diagnosis is possible. Thus a multiple test cycle is required. If, however, the available knowledge about the physical properties of the PCB under test is used, some constraints can be put on the occurrences of faults and this will relax the conclusion somewhat.

CLUSTER TESTING

Although Boundary-Scan testing has evolved enormously it is believed that for many years to come PCBs will contain both Boundary-Scan and conventional components *without* BST facilities. It is expected that most ASICs and only a moderate number of commercial devices like transceivers and some microprocessors will be provided with BST facilities. For the low cost end of the available components it may take many years before the extra design effort is worthwhile to include BST, if it will happen ever. Therefore, most PCBs will contain mixed-technology logic well into the future.

A mixed-technology board requires a hybrid test strategy. The conventional logic clusters will be tested on the conventional way, with bed-of-nails fixtures applying parallel test vectors for in-circuit test techniques. The Boundary-Scan components and their interconnecting nets will use serial test patterns which are shifted through the BST chain. This section addresses some of the test issues in relation to these 'mixed' boards. Figure 5-7 shows a sample circuit containing both BST and non-BST logic.

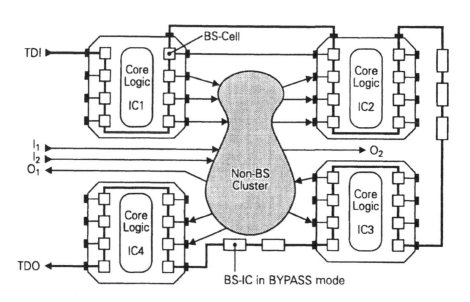

Fig. 5-7   Circuit containing BST and Non-BST logic

As can be seen, a cluster may have its inputs and outputs connected to Boundary-Scan ICs, to other circuitry or to board connectors. Some other Boundary-Scan ICs are indicated which are put into the BYPASS mode while this particular cluster is being tested.

A convenient way to apply test signals to the cluster and observing its responses is to consider the driving and sensing nodes of the connected BST ICs as virtual ATE tester nails. The test stimulus for the cluster is loaded through the TDI-TDO path into the relevant BS output cells. The responses are captured in BS input cells and shifted out for diagnosis.

Boundary-Scan testing slows down the cluster testing in various ways. Through conventional ATE channels, the test signals can be dynamically formatted, for example returned to zero or set to one. Such type of actions through BST can only be done in several test cycles, each one shifted in sequentially. From a designer's point of view, the most convenient way to produce the cluster test patterns remains still the one which assumes that the tester applies and detects parallel test vectors. But for Boundary-Scan testing sequential test patterns are required, meaning that a tester should include a serializer which transforms the parallel vectors to sequential/serial vectors. A simple architecture that adds flexible vector sequencing and control capabilities to a general purpose digital tester is proposed in [36]. For analysis of the test results, the tester software may have to deserialise captured test data. However, the serialization process will inherently increase the amount of test data.

A cluster may be connected to only a few out of a couple of hundreds pins of the respective Boundary-Scan (BS) ICs, of which each pin is provided with a Boundary-Scan Cells . Further, as can be seen in figure 5-7, the Boundary-Scan chain also includes a number of BS ICs which are not used while the cluster is tested. Although these BS ICs are put into the BYPASS mode, they require some extra test clock cycles to pass the relevant test data through the chain into the BS cell concerned. From this observation it will be clear that only a small part (2 to 3%) of the shift-register elements in the BS chain deliver or receive cluster signals. But every shift-register element in the BS chain must carry some logical value for the test.

It is obvious that part of the cluster test is still to be done in the conventional way. However, the board designer can contribute significantly to this test strategy by taking care in advance of the cluster's properties (DFT). For example the designer could split up the cluster into more and simpler parts. These parts then may have more connections to the Boundary-Scan circuitry and have a reduced sequential depth which encourages the automatic generation of the test patterns. Moreover, clusters may be subdivided into various *types* of cluster, i.e. memory array clusters (see next section), single device clusters (glue logic) and random logic clusters, such as programmable logic devices (PLDs). For each type of cluster the optimal/dedicated test pattern can then be applied, like counting sequences and walking ones or zeros, which limits the test length. When the limited test speed for

BST causes problems for dynamic parts, the designer could provide keep-alive signals through the conventional ATE tester nails.

Finally, next to the test problems associated with the serialization and partition of the cluster test patterns, a general concern is the automated diagnosis of cluster faults. The tester must 'know' the PCB topology in order to select the meaningful test data out of the entire background data stream.

Impact on Fault Detection

Compared to full Boundary-Scan boards, some particular problems are encountered with the testing of boards that contain both BS and non-BS ICs. For one it must be realized that the board is powered up during BST and that the non-BST clusters can not be observed and fully controlled through the Boundary-Scan cells. So, additional to Boundary-Scan control mechanism, conventional in-circuit access to the non-BS cluster nodes remains necessary for testing.

Referring back to *Fig. 5-7* it can be seen that the non-BST cluster may react unpredictably on signals arriving from IC1 or that the input signals I1 and I2 may cause an unexpected reaction on the sensing node of IC3 for example. The cluster may also contain a free running clock for internal cluster logic that is otherwise not to control, which makes any form of test generation very difficult. To develop the new test methodology two constraints need to be kept in mind:

• the conventional cluster is not fully controllable through the BS path and

• the logic within the conventional cluster can possibly not be initialized and hence the cluster will behave unpredictably.

This may, but need not, cause an unrepeatable failing test if only the powering up of the PCB for BST causes a new, irreparable fault. It is clear that the designer can avoid such draw-backs beforehand, if Design For Testability (DFT) is applied. A cluster can be subdivided into smaller pieces so as to provide more visibility and testability. Some or may be all cluster parts could also be made a BST compatible ASIC.

Finally it should be remembered that in an electronic circuit diagram nodes may be assigned which are neither a Boundary-Scan cell nor a node accessible to ATE tester nails. For testing purposes these nodes are invisible and called inaccessible. Faults that include such nodes could, however, confuse the test algorithms and hence the test results.

These observations have led to a mixed test methodology consisting of four distinct test steps [37]. Each step covers part of the occurring faults, whereas all steps together will provide sufficient information for a final and complete diagnosis of the fault. Though these test steps can be taken in any order, it simplifies the diagnosis when they are performed in the sequence as given here.

1. Conventional Short Test

A test for shorts between only those nodes which are accessible for the tester nails. This test is safe to do first, because it can detect many shorts which could damage PCB components if the board was powered up for a Boundary-Scan test. When faults are discovered, testing interrupted and faults could be repaired before continuing with the next test step.

Note that if the bed-of-nails fixture was also designed to access the BS path between TDI and TDO, the shorts between the Boundary-Scan chain and the non-BST nodes could be detected. But this is only a theoretical option, because the IEEE 1149.1 standard has been introduced primarily to test the interconnects on the PCB.

2. Boundary-Scan Integrity Test

The benefit and the use of this test of the Boundary-Scan path has been discussed in testing the integrity of the Boundary-Scan test chain, earlier in this chapter.

3. Interactions Test

This part of the test sequence checks on shorts between the Boundary-Scan nodes and other nodes (non-BST) which are only accessible for the tester nails. It also checks on opens between a tester nail and the input pin of a Boundary-Scan IC.

This test is also used to check the behaviour of some Boundary-Scan nodes when they are being back-driven by an opposite voltage from the tester nails. This simulates the practical case for boards that contain conventional clusters which interact with Boundary-Scan ICs.

The choice of the test patterns and the diagnostic considerations remain as described earlier in this chapter.

4. Boundary-Scan Interconnect Test

These are the 'normal' tests as described so far in this book. Any further discussion of this test is omitted here.

Conclusion

The presence of inaccessible nodes within clusters needs some extra attention. Shorts between such nodes will probably not be detected or can not be diagnosed and will only become apparent during the functional test. Shorts between inaccessible and accessible nodes will be detected probably in either the third or the fourth test step mentioned above. Partioning the cluster into various types of circuit (DFT) will alleviate these problems, because the cluster becomes more transparent and adaptive to dedicated test sequences for each type. But above all, the expense of in-circuit testers can be greatly reduced by applying DFT and introducing BST circuits wherever possible. The savings in test costs during the PCB's life cycle are indicated in chapter 1 and shining examples are given in chapter 6.

Interconnect Test of Memory Clusters

Cluster tests using Boundary-Scan testing can result in long test times because of the serial nature of the Boundary-Scan Test access and the generally high number of test patterns required. This section shows an approach to the interconnect test of *memory* clusters in which special attention is paid to obtain a small number of test patterns, so making memory interconnect tests feasible for Boundary-Scan. The proposed test algorithm provides for an optimal fault detection and diagnosis of stuck-at and bridging faults of memory interconnects [38].

To develop the proper test algorithm the following aspects of memory clusters are considered. A memory cluster may either consist of a single memory chip or of various different chips. Different types of memory may be present; both read only memory (ROM) chips and random access memory (RAM). The latter type of memories are subdivided into dynamic and static RAMs. Further, the memory cells of dynamic RAMs have to be refreshed frequently. This requires special methods, since it is not feasible to refresh the memory by shifting data through the Boundary-Scan chain. This forms a separate constraint with respect to testing.

In the following discussions we consider the interconnection tests for single a RAM followed by those for a ROM. The final paragraph describes some test considerations for memory clusters consisting of several chips.

The interconnection tests involve the address lines, the data lines and the control lines.

As to the *fault model*, only the single stuck-at fault model is assumed although multiple occurrences of it in different nets are considered. Only bridging faults between two nets are assumed but here again, multiple occurrences of them are also considered. A wired AND is assumed for the bridging fault, but adapting the test could allow testing for wired OR shorts. Non-deterministic or undefined behaviour is not considered. For Random Access Memories a chip with n address lines and m data lines is assumed.

Random Access Memory

In order to give an idea how a memory READ and a memory WRITE cycle can be emulated by applying a Boundary-Scan test, the timing diagram in figure 5-8 is used.

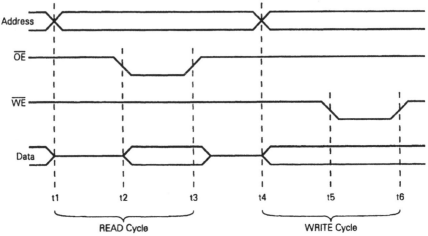

Fig. 5-8   Timing of RAM accesses

In this figure OE denotes Output Enable and WE is Write Enable. Often the Memory Read and Memory Write signals are complementary on a WE line. Further it is assumed that the OE and the WE signals do not change at the same moment as the Address signals. This would cause additional overhead on the access if this is done through Boundary-Scan test devices. As can be seen from the diagram, the READ cycle consists of three steps:

t1:   the required address is driven to the address lines,

t2:   the output is enabled to write data to the data bus,

t3:   the data is read and the output is disabled again.

As an example a memory chip is considered with 8 address lines and 4 data lines. A READ operation of address '11001011' is achieved by the test vectors 1, 2 and 3 in table 5-12. The data stored at this address is '0011'. The WRITE operation is performed in the same way by test vectors 4, 5 and 6: the data word '1101' is written to address '10111000'.

Table 5-12   READ and WRITE vectors at a RAM

No.	Address	OE	WE	CS	D_{out}	D_{in}
1.	1100 1011	1	1	0	TTTT	NNNN
2.	1100 1011	0	1	0	TTTT	0011
3.	1100 1011	1	1	0	TTTT	NNNN
4.	1011 1000	1	1	0	1101	NNNN
5.	1011 1000	1	0	0	1101	NNNN
6.	1011 1000	1	1	0	1101	NNNN

During a READ operation the Boundary-Scan cells that control the data line are tri-stated, in the table denoted as 'T', in order to avoid bus collisions. The results are captured by the cells that observe the data lines. The results are unknown, in the table denoted as 'N', unless OE is active. Because only one chip is assumed here, the Chip Select (CS) line can be kept active. This will be selectively set if more chips in the cluster are to be updated. This procedure proves that a READ and WRITE cycle can be emulated through Boundary-Scan cells in a chain.

The test time 't' in the following paragraphs refers to the formula:

$$t = L \cdot (w \cdot c_W + r \cdot c_R) / f$$

where

L   = the length of the test vector,
w   = the number of WRITE operations,
r   = the number of READ operations,
c_W = the number of test vectors required for the WRITE operation,
c_R = the number of test vectors required for the READ operations and
f   = the test clock frequency.

Address Line Faults

Faults on address and/or data lines cause a change on some addresses or a transformation of some data words. If a WRITE operation is applied, a change of an address causes another memory position to be written to. If the overwritten data position is known, this can be detected, by an algorithm that is basically described in [39]. The principle is that sequentially two adjacent address lines are written to and the results at the first line are compared with the expected values. The only care to be taken is to the first address line of (binary) 0, which does not change if a bridging fault occurs. If only one address line is kept at 1 and all other address and data lines are kept at 0, then a bridging fault or a stuck-at 0 fault on this line will cause a change to address 0. Therefore, for the first WRITE operation, the initial contents of address 0 must be known and not be 0 in order to detect the fault-free situation. If the above mentioned test algorithm is used for each address line separately, all possible faults on address lines are detected. Applying the walking one sequence to test all n address lines, only n such test sequences are needed, which makes it suitable for Boundary-Scan testing.

Data Line Faults

Fault on data lines cause data transformation, which can be tested by applying a WRITE and a READ emulation respectively. Bridging faults between data lines and address lines cause a stuck-at 1 behaviour as reasoned above. Testing data lines for stuck-at and bridging faults is quite similar to testing for wiring interconnects. For m data lines both for WRITE and for READ operations the number of test operations is: $p = \lceil \log_2(m+2) \rceil$.

For the detection and diagnosis covering stuck-at and bridging faults, a test pattern occurrence of both 01 and 10 at each pair of data lines is applied [40]. It turns out that this test sequence for data lines is a subset of the test for address lines, mentioned above. Therefore it is possible to test for defects on address and data lines simultaneously, which means a major decrease of required test patterns. This, again, makes the test even more 'BST friendly'.

Using the above mentioned formulas the following example may be enlightening. A memory with 20 address lines and 8 data lines is taken for the example. For the 8 data lines $p = 4$. In [38] it is shown that for the above tests only $n+p$ WRITE actions and p READ actions are sufficient for *only* detection, making it 24 WRITE and 4 READ operations. If the Boundary-Scan test vector length is taken at 50 and the clock frequency at 10 MHz, then a Boundary-Scan test takes 0.4 ms.

For diagnosis the test must provide more 'complete' data, requiring a larger number of test patterns. But even then, for the same example, it is shown that for fault detection *and* diagnosis the total test time increases to 0.9 ms.

Both times are a kind of worst case, taking into account that the number of test actions varies due to the question whether p is larger or smaller than n.

Control Lines

These lines comprise the enabling lines, like read enable, write enable or chip select. Faults on these line may lead to reading random data words or even one data word throughout the whole test. It is not always possible to uniquely determine the cause of such detected erroneous data.

Read Only Memory

Fault detection and diagnosis of interconnection faults to a memory cluster requires much less test application time than reading the contents of the ROM. If p is the sum of the number of data and address lines, then the number of single stuck-at faults is $2p$ and the number of bridging faults is $p(p-1)/2$ (see above sections in this chapter). An address is read only if at least one fault can be detected by this READ operation. Therefore $p(p-1)/2$ addresses suffice to detect bridging faults and maximum p addresses suffice for the remaining stuck-at faults. Thus, for fault *detection* only, the required number of READ operations becomes $p(p+1)/2$. If diagnosis is also required, $p(p-1)$ address have to be read to detect and locate the bridging faults and maximum $2p$ addresses are required to detect and diagnose all single stuck-at faults. So for detection *and* diagnosis a maximum of $p(p+1)$ addresses have to be read. If for example a memory chip has 20 address lines and 8 data lines, then 56 single stuck-at faults and 756 bridging faults are possible. So in total 812 test vectors are sufficient for detection and full diagnosis of all single interconnect faults. Evidently, this is no problem for Boundary-Scan testing.

This number may, however, rise because the obtained fault coverage and diagnosis depends also on the contents present in the ROM. A fault on an address or data line can only be detected if the data value is changed, so that the detected value differs from the expected value. The ROM's data can cause a masking of the fault if aliasing, confounding or oscillating effects occur during testing (as previously described in this chapter). Addresses are selected in order to detect ROM interconnect faults. The contents of the ROM are therefore searched for appropriate addresses and corresponding data values to detect specific interconnect faults. Each address is tested for aliasing, confounding and oscillations. If the diagnostic resolution has to be high in order to avoid confounding results, a fault dictionary is required. Still, which patterns have to be chosen depends on the data present on the other locations of the ROM. If pin positions are provided, the majority of the bridging faults are not likely to exist and only $(m+n-1)$ possible bridging faults remain.

For the same example, 20 address and 8 data lines, only 27 fault combinations remain, which further reduce the number of required READ cycles.

Memory Clusters

A memory cluster consists of various memory chips connected to the same address and data bus. Each chip has to be tested separately for interconnect faults, the cluster contains both ROM and RAM chips. The number of test patterns for bridging faults is reduced significantly, because bridging faults occur for all chips working on the same bus. If data and/or address lines are multiplexed within the chip, the test size is also reduced because multiplexed lines need only to be tested once.

When an external latch is used, a single fault at multiplexed address and/or data lines occurs as a multiple fault, caused by the multiplexing process. This will significantly increase the test generation complexity, because, other than the assumed single interconnect faults, possible multiple faults have to be taken into account.

Conclusion

With the development of a minimum set of test vectors as indicated above, it appears quite possible to test a memory cluster for interconnection faults by means of Boundary-Scan Test. The examples given in [38] show BS tests for READ and WRITE cycles that detect *and* diagnose interconnect faults in less than a millisecond.

ARCHITECTURE OF A BOUNDARY-SCAN TEST FLOW

Figure 5-9 depicts a general architecture of a Boundary-Scan test flow.

The dataflow diagram in this figure can be partioned into three phases:

- the test pattern preparation and (automatic) generation,

- the test execution phase in which the Unit Under Test (UUT) is being tested,

- the result analysis or diagnosis.

- *Test Pattern Generation*
  The input for the first phase usually comes from a CAD/CAE environment. The three right hand files at the top in figure 5-9 may result from a Boundary-Scan Register Compiler running on a CAD station, as described in chapter 4. Such a compiler uses a design description and a technology library as input to generate the netlist described in EDIF [23], the components' data sheet information in BSDL and some specific assignment files (compare figure 4-5). The compilation of these files is called the Net Assignment File.

This file is the input for the actual test pattern generator, called ATPG (Automatic Test Pattern Generator) or BTPG™ (Boundary-scan Test Pattern Generator [41]). A typical tester generates test vectors on user demands. For example the Boundary-Scan infra-structure can be tested on integrity through either the BYPASS-register, the ID-register or the complete Boundary-Scan

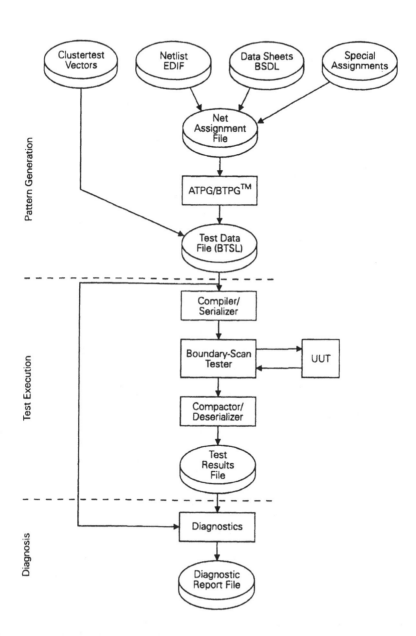

Fig. 5-9   Block diagram of a Boundary-Scan tester

chain. Next to the integrity test the BS tester should test at least the PCB's interconnecting nets, cluster tests and component-internal tests (RUNBIST). The interconnect tests comprise bridging faults, opens, stuck-at faults and extended stuck-at faults for bus structures.

Typically, PCBs have a combination of BS and non-BS devices. Therefore externally generated parallel test vectors need to be imported into the test vector set. The resulting file then is the Test Data File describing the full test specification which usually does not come in a tester dependent format.

› *Test Execution*
The Test Data File is converted and serialized to suit the board tester. The resulting file should obey the IEEE 1149.1 BST standard protocols. Now, the UUT can be connected to the actual tester hardware and the software can be run to make things happen, i.e. perform testing. The test hardware may consist of a simple PC extended with a Boundary-Scan test interface box or of a sophisticated logic analyzer which is upgraded with a unit capable of handling the BST software. The output from the UUT is collected back into the tester, where the background patterns are removed (compacting) and the vectors are deserialized again to the usual parallel format. A memory image of the test result may be collected on disk in a test results file for later processing or it may be fed immediately to the diagnosing software for on-line diagnosis.

› *Analysis and Diagnosis*
Obviously, for diagnosis the test results have to be compared to the expected values, for which the required information is abstracted from the test data file. Diagnostic reporting may occur at various levels.

- A fault summary is generated stating per fault the fault type, the net name and the detected values against the expected ones.

- Additional to the summary more information can be supplied concerning the faulty net's connections, control signals and Boundary-Scan bit positions.

- Bit information can be displayed, uninterrupted by the tester, showing the bit information for all failing test vectors.

Apart from the level of diagnosis, the format of the screen representation may vary. Broadly speaking two formats can be distinguished: the truth table format and the Boundary-Scan interpretation. For the truth table format various sub-screens may be available for reporting cluster faults, interconnect faults, capture and identity faults. The faulty bits in the test vectors may be displayed in reversed video or highlighted. The Boundary-Scan screen may display the various fault types in table format on the screen, for example for ID faults, stuck-at and Bridging fault, Net bridging faults or list the Error Vectors. The next pages show two examples of the BST representation. In the last chapter the truth table representation is shown in the section where a PC based BS tester is described as an example of a very cheap tester.

Finally it should be noted that the diagnostic results can be collected in a file for later discussions or for legal purposes.

As an example figures 5-10 and 5-11 show two screens with diagnostic results displayed from a Diagnostic Report File such as in figure 5-9.

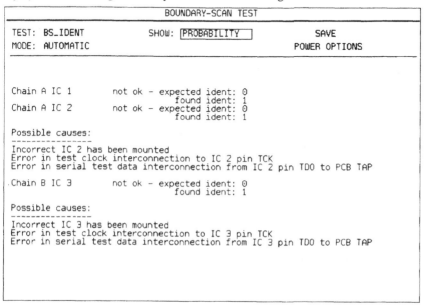

Fig. 5-10   Screen display showing identification faults

Serializer Considerations

Given the facts that in the above Boundary-Scan test flow the test patterns are in parallel format (BTSL) and that designers prefer to think in parallel test vectors, it will be clear that the process of sequential pattern generation needed for BST requires extra attention.

Referring back to figure 5-9 it should be noted that the serializing process is only needed to bring the test data into the appropriate Boundary-Scan cells. The testing itself on the PCB is actually a *parallel* test action. The next shifting action brings out the test results which are by the shifting nature in serial format. For further detection and diagnosis these test results are handled in parallel format again.

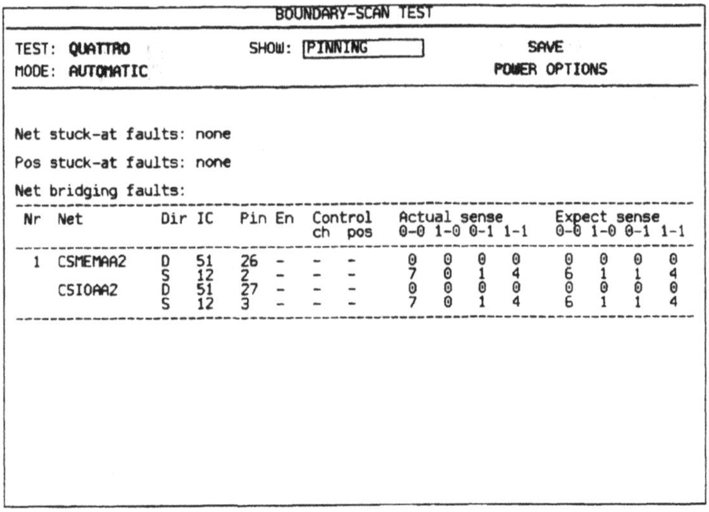

```
                          BOUNDARY-SCAN TEST
 ┌──────────────────────────────────────────────────────────────────────┐
 │ TEST: QUATTRO              SHOW: [PINNING      ]        SAVE            │
 │ MODE: AUTOMATIC                                  POWER OPTIONS         │
 ├──────────────────────────────────────────────────────────────────────┤
 │                                                                        │
 │ Net stuck-at faults: none                                              │
 │                                                                        │
 │ Pos stuck-at faults: none                                              │
 │                                                                        │
 │ Net bridging faults:                                                   │
```

Nr	Net	Dir	IC	Pin	En	Control		Actual sense				Expect sense			
						ch	pos	0-0	1-0	0-1	1-1	0-0	1-0	0-1	1-1
1	CSMEMAA2	D	51	26	-	-	-	0	0	0	0	0	0	0	0
		S	12	2	-	-	-	7	0	1	4	6	1	1	4
	CSIOAA2	D	51	27	-	-	-	0	0	0	0	0	0	0	0
		S	12	3	-	-	-	7	0	1	4	6	1	1	4

Fig. 5-11   Screen display showing net bridging faults

In order to transform the test stimuli into a serial format as economically as possible, various techniques have evolved. To start with, it should be remembered that for BST *two* operations are involved to present *one* scan vector: first the test stimuli must be shifted into the BS chain and secondly the responses must be shifted out. During the second operation the stimuli for the next test are shifted in again, etc. Then, in particular when dealing with cluster testing, the topology of the PCB must be known, so that the few percent of relevant cluster test data can be situated at the proper place in the large bit stream.

Before feeding the test stimuli into the TDI pin of the PCB's TAP, the serializer must have specified the state of the TMS for each TCK cycle. The resulting response data are concealed in a bit stream of irrelevant data and are of an unknown state: '1', '0' or 'X', for which two bits are required for detection. Thus for generation and detection of the test data the serializer must provide the following data bits for each clock cycle:

- one bit of test data to the TDI,

- one bit of control data for the TMS setting,

- two response-data bits for the test hardware to determine the signal states '1', '0' or 'X' at the TDO pin.

Now the tester is supplied with all relevant PCB data: the topology of both the cluster and the Boundary-Scan path, the required cluster test data and the value of the irrelevant background data bits. Finally, for the serialization process several options are available [42].

The options for serialization are illustrated here with the aid of an example.
For the sample cluster it is assumed that it has 40 inputs and 30 outputs connected to the Boundary-Scan ICs, that 10,000 test patterns are needed and that per pattern one bit for the input ('0' or '1') and two bits for the output ('0', '1' or 'X') are required.

1. Run-Time Software Serialization

Only the cluster test data are stored. The serialization process takes place each time a PCB is inserted into the tester. This strategy minimizes the data storage requirements, which then become:

$$(40 \cdot 1 \text{ bit/input} + 30 \cdot 2 \text{ bits/output}) \cdot 10,000 \text{ patterns} = 10^6 \text{ bits}.$$

The size of the (compressed) topology data and background pattern are about of the same magnitude and they are negligible, a few thousand bits in total. The only disadvantage of this strategy is the long test time when the serializing process is run on a general purpose computer (PC). Then the Boundary-Scan TAP controller and scan path hardware is running some orders of magnitude slower than should be.

2. One-Time Serialization With Run-Time Access

The serialized bit stream is stored in the tester and is accessed each time a PCB is tested. Assuming that the Boundary-Scan path is 1000 clocks long and that four data bits are required for each clock cycle, then the stream of test patterns for the sample cluster adds up to:

$$1000 \text{ clocks} \cdot 4 \text{ bits/clock} \cdot 10,000 \text{ patterns} = 4 \cdot 10^7 \text{ bits}.$$

Even for the modern testers this amount of data takes an inefficiently long time to load it from the tester's disk unit.

3. One-Time serialization With Large Scan Buffer

A large scan buffer is directly accessible for the ATE channels. The PCBs can be tested at scan rates that are only limited by the Boundary-Scan path hardware and the tester itself. But the buffer should be able to hold the test program of the entire board. If such a board comprises say 10 of the abovementioned sample clusters, then it will be clear that this strategy will lack any efficiency.

The only feasible option is to test a large number of the same PCB. But ASIC loaded boards are produced frequently in small series and must be available on shortest notice (JIT manufacturing) for competitive reasons. So even this option is unlikely to be met.

4. Run-Time Hardware Serialization

A dedicated hardware processor executes the mixing of the cluster patterns with the background patterns and takes care of the topological data for both stimulus and response vectors while each board is being tested. This strategy eliminates both the long run times and the inefficient storage requirements. The scan rates are only limited by the Boundary-Scan path hardware and the tester itself.

5. Serialization of Algorithmic Patterns

So far it is assumed that the cluster tester received the serialized test data as a linear set of patterns, or as a truth table, which may come as a result of a logic simulation from the designer. A special type of cluster is a configuration of memory devices. These are tested using algorithmically generated patterns, which can be described in a terse manner, rather than by very long truth tables.

ROM testing is performed by sequentially reading every address and applying the response data to a Multiple Input Shift Register (MISR) after being extracted from the output bit stream. So a process of comparison of the captured data with the expected data (as with the previous cluster data) is not applicable.

RAM testing is typically performed by a 'march' routine which accesses each location for writing and reading the binary values '0' and '1' to-and-fro. The serializer creates all stimulus and response values and testing takes place in hardware as in the initial cluster mode.

In *conclusion* of the discussed serialization options it is seen that, when random clusters are to be tested in a Boundary-Scan environment, designing of dedicated test hardware yields the most efficient test pattern generation. Algorithmically generated test patterns are best used for memory cluster testing.

Chapter 6

MANAGEMENT ASPECTS

After an overview of the birth of the IEEE 1149.1 Standard, this chapter describes the measures to be taken in order to introduce Boundary-Scan testing into a company. Points of attention are impacts on the product's life cycle, product reliability and quality and —most importantly— cost calculations.

COMING TO THE STANDARD

In chapter 1 of this book the need for new electronic test methods as a result of the ongoing miniaturization is discussed. Driving sources for miniaturization were coming from the need for high performance systems such as supercomputers, mainframes, workstations and the introduction of portable small sized consumer products like hand held digital phones, car phones, laptop PCs, notebooks etc.

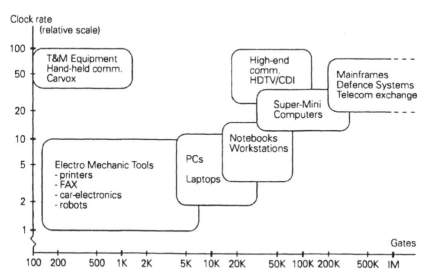

Fig. 6-1   Relation between complexity and speed of electronic systems

Figure 6-1 indicates the relationship between these systems in terms of complexity (number of gates) and internally used electronic working speeds (clock rate).

173

The growing demands for these highly sophisticated products has led to much functionality in small sized packages. The electronic components used have become highly integrated ICs, VHSICs and Multi Chip Modules mounted in units with over 500 pins at pitches of 0.3 mm or less. Moreover, there is a continuous drive to improve the reliability of highly complex systems as used in telecommunication, high-end computers in civil and defence applications. This drive has led to an increased awareness of design for testability (DFT), a design method in which the testability is provided right from the start of a product design.

Within just a few years, the price of the automatic test equipment suitable for these products has increased from $100,000 to more than $1.5 Million.

In the factories, the PCB test preparation costs and lead times for software and bed-of-nails fixtures grew at the same rate. Today, the management of production lines may even refuse to build extremely dense fixtures for too complicated PCBs.

It became clear to all test engineers in the world that this process could never continue and had to come to an end. Therefore test engineers at Philips Electronics took the lead and contacted test engineers of other electronic equipment makers in Europe to set up a committee to tackle this test technology problem. The first session took place in November 1985 and right from the beginning a consensus was reached about the problem. And what's more, a unanimous willingness to cooperate in solving the test problem existed. The result was that the Joint European Test Action Group (JETAG) was formed and appointments were made for further meetings. The chair was held by Harry Bleekers of Philips Electronics. To the participating companies belonged British Telecom, Bull, Ericsson, Nixdorf, ITT (now Alcatel), Siemens, Thomson and others.

It was agreed that the following subjects should be treated.

- Define a new test methodology for PCBs which can cope with the current test problems as well as those expected in the future. Therefore, the methodology should remain 'future-proof'.

- Specify the test method for both components and PCBs. It is important to achieve compatibility, i.e. components of brand A should be able to 'talk' to components of brand B on the PCB. This requirement can only be assured if the test method is fully specified in all details.

- Promote the new methodology to other companies, including IC suppliers, by means of conferences, workshops, etc. There existed a vicious circle. The IC vendors were not willing to implement new test provisions into standard ICs if there is no market demand and the PCB designers were not introducing the new test standard if there are no ICs available for it. This situation has been overcome by promoting very heavily the advantages of Boundary-Scan Test in the area of the ASIC-technology: here the IC and the PCB designs are created in one environment.

• Submit the new methodology for standardization to a world-wide recognized authority such as the IEEE. It is obvious that a standard should be controlled by an independent body and not by a company because stability of a standard is vital. It is fatal if changes to a standard come out and make previous investments obsolete.

The interest in JETAG grew and North American companies joined the JETAG, which then was not a purely European initiative; hence the name JETAG became JTAG. Amongst the North American companies were AT&T, DEC, IBM and Texas Instruments. JTAG considered it its goal to encourage IC suppliers and electronic test equipment makers to introduce the envisaged test methodology into their companies, thus to apply it in their ICs and testers.

That the JTAG was successful in its sustained effort (16 JTAG meetings were held) can be deduced from the letters of endorsement, received from almost all leading companies. A few of all the received statements are reproduced here.

AT&T

We will begin the process of increasing vendor awareness of the industry need for implementing the JTAG standard - as we are beginning to do in AT&T ICs.

Laurence C. Steifert, Vice President Engineering, Manufacturing and Production Planning

IBM

We are happy to participate in the JTAG efforts to generate and establish such standards for Boundary-Scan.

J.C. McGroddy, Assistant General Manager for Technology Products

PHILIPS ELECTRONICS

Philips strongly supports and asks for implementation of Boundary-Scan in commercial digital VLSI-circuits in order to achieve their high quality objectives.

Ir. C. Kooij, Managing Director of Centre for Manufacturing Technology

TEXAS INSTRUMENTS

The goals and objectives of JTAG are consistent with TI's philosophy of providing products that are both functional and testable, thereby reducing the overall cost of ownership.

George H. Heilmeier, Senior Vice-President and Chief technical Officer Corporate R&D

These endorsements implicitly confirmed the following facts.

- The managements of the participating companies became aware what was happening in their own company.

- The JTAG representatives received support from their respective managements to travel in order to attend the JTAG meetings.

- In relation to the IC vendors it was demonstrated that 'companies' were requiring the testability provisions and not just a few test engineers.

In 1987 the JTAG architecture version 1.0 for loaded-board testing was proposed following a paper from the Philips Research Labs [43], succeeded in 1988 by version 2.0. The latter version was submitted as a standard architecture to the IEEE Computer Society's Test Technology Committee. The JTAG activities were approved when in February 1990 the IEEE Standards organization introduced the IEEE Standard 1149.1-1990 "Standard Test Access Port and Boundary-Scan Architecture" [1].

The last JTAG was held on March 9, 1990 to cease the activity because the objectives were met. The regular maintenance for the standard was now in the hands of the IEEE 1149.1 working group. In August 1990 the American National Standards Institute (ANSI) also recognized this standard.

Effects of Boundary-Scan Testing

The utilization of Boundary-Scan Test requires provisions for testing to be made in the design phase of ICs which, later on, are used for testing the PCBs containing those ICs. As such the introduction of BST encourages the design-for-test (DFT) philosophy which, on the other hand, may add some costs to the original designs. However, the revenues are earned back many times over, due mainly to the simplified testing methods in the factory and in field service. The costs of both automatic test equipment and test set-up times are reduced drastically.

Reduction in production time is not the only place of cost-saving. BST also reduces prototype debug time, thus improving the overall time-to-market for a product. Table 6-1 gives an indicative overview of the advantage of applying Boundary-Scan Test in an organization. In this table a '+' denotes a positive effect and a '−' is negative effect. It is clear that the overall effect can significantly improve a company's market position, maybe resulting in a market leadership, which is *always* profitable. The following sections describe worked out examples.

Notice that BST adds test facilities to functional logic. This means that testing becomes independent of implementation technologies, i.e. it can be implemented into silicon, ECL, GaAs, onto PCBs, MCMs, systems, etc. So in this respect *BST is technology independent.*

Table 6-1 Relative indicators for BST applications

Subject	Indicator
Design engineering	
- IC design	−
- Reusable test vectors	+
- Test program generation	++
- Board design time	−
- Board prototyping	++
Manufacture	
- Test pattern generation	++
- Test time	++
- Material costs	−
- Diagnosis	++
- Repair	+
- Retest	+
- Test equipment costs	++
Commissioning	
- Diagnosis, replacement and repair	++
Field maintenance	
- Diagnosis	++
- Replacement and repair	+
Marketing	
- Time-to-market	++

MANAGEMENT ROLE

Since the technical side of the test problem caused by technological changes (miniaturization) has been solved after approval of the IEEE Std 1149.1, the introduction of BST in a company has become a managerial issue. As BST covers the whole product's life cycle (design, manufacture and service), sub-optimization of the budget in the various departments will hamper the integral approach for cost reduction. Here the management should take the lead.

To demonstrate the integral approach, consider figure 6-2.

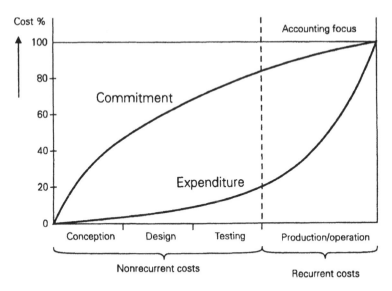

Fig. 6-2   Relation between commitment and expenditure
Source: Electronic Design, March 28, 1989.

One interesting aspect is the fact that the major part of the costs over the life cycle of a product is spent in the manufacturing and operational phase. These costs are determined by decisions taken in the *design* phase of a product. Thus the only way of later reducing costs is through complete redesign.

The role of the management is to recognize these affects. Boundary-Scan Test reduces the manufacturing costs since it implies a Design For Testability (DFT) method. The expenditures in the design phase are minimal. Moreover, the decisions for expenditures in the design phase concern the rather short term costs as opposed to long term investments. The next sections elaborate upon various management aspects which help to increase the overall profit on new electronic products.

Introduce Concurrent Engineering

It can be stated that the introduction of Boundary-Scan Test favours the introduction of concurrent engineering. It should be made clear that concurrent engineering can be introduced without BST, but BST further increases the advantage. In any case, the main reason to introduce concurrent engineering is, of course, to get the right product, at the right time, for the right price, onto the market.

As opposed to sequential engineering, concurrent engineering uses a product development cycle in which various manufacturing disciplines work *simultaneously* together on the design of a product. These disciplines may consist of design, test, manufacturing, marketing and field service.

For concurrent engineering a project team is put together with engineers from these disciplines. It is evident that such a team is continuously supported by the management. When the team members meet at regular intervals, say weekly, the internal communication will prevent redesigns due to design errors or due to insufficient design verification. As a result, a lot of costs are saved during production, earning back many fold the extra time spent in the design phase. If a redesign is not unavoidable, it should be done so as to fit the design into the current manufacturing process rather than adjust this process to the new design. In this way unnecessary costs are avoided. Finally, designers are encouraged to cooperate when they are involved in the new product right from the beginning, as soon as the proposals are put forward by the marketing department. As a matter of fact, this is also quality improvement because, due to the encouragement, the designer may put more added value to the product.

From a social point of view however, it may take some effort to introduce the change in culture. Initial support of the project team may be needed to allow the members to communicate with each other. The brought-in proposals should be relevant, in time, accurate and put forward with an intention of improvement and not as a kind of punishment on the others.

The project team may be encouraged to focus also on *cost related* subjects of the development cycle and be alert to cost *savings*.

* *Design costs*
  These costs may consist of bought-in capital goods (test and measuring equipment, machines) and their amortization periods, component costs, hardware and software design costs, test program generation and design verification as part of the integral test-plan.

* *Manufacturing costs*
  These include initial costs for machinery and equipment for board assembly and board testing (fixtures) and their amortization periods, probably additional test program generation, recurrent costs for board/system assembly testing and repairing and the shipment costs.

* *Quality costs*
  Testing improves the quality output but in fact it does *not* add value to the product. Therefore, in this respect, quality *costs* money; but it *saves more* money. Nevertheless, time has to be spent on component, board and system inspection, monitoring the manufacturing process for improvements and an

overall analysis of the development and production cycle. The analysis may even include the field service costs.

- *Service costs*
  Depending on the product this might include customer training, product replacement, guarantee costs, repair costs either at the customer's site or in the factory, stocking of spare components, subassemblies or products and travel expenses.

Figure 6-3 shows a cost comparison of two product development cycles.

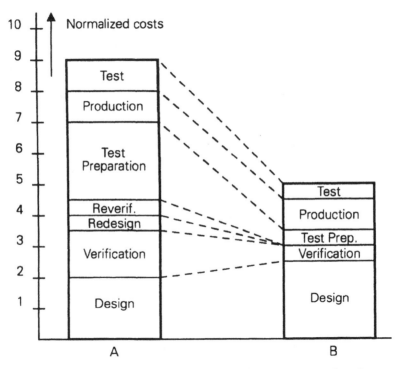

Fig. 6-3   Costs bars for (A) sequential and (B) concurrent engineering

The costs are given as "normalized" to indicate the comparative character of the figure. It can be seen clearly that in the concurrent engineering bar some extra costs are spent on design which are recuperated greatly later on. Redesign and reverification present in the sequential cost bar have even vanished completely as a result of the mutual communications in the project team. The design for testability has reduced the test preparation time significantly. The final test cost is reduced by about a half.

As stated above, Boundary-Scan Test will definitely help to introduce concurrent engineering. The next example will show how even concurrent engineering is improved by the introduction of BST, thus reducing production costs even more.

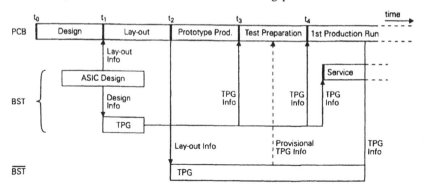

Fig. 6-4 PCB design cycle with and without BST

In figure 6-4 the design cycle of the PCB is given along a time axis, marked with points t_n, indicating start and finishing points for each phase in the design cycle. The lowest line is indicated with $\overline{BST}$, which denotes the test pattern generation (TPG) in the case that no BST is applied in the PCB design.

If BST is applied the working method is as follows. As can be seen, shortly after the start of the PCB design (t_0) a start is made with the ASIC(s) design(s). At the instant t_1 the ASIC's pin-out has been fixed and consequently the Boundary-Scan path is defined and the cells are generated, which can be done automatically as described in chapter 4. The design of the ASIC core logic does not need to be 100% ready at that instant, but the design of the PCB layout can be started because the ASIC's pin definitions are ready. Concurrently, the test patterns can be generated (TPG), which are easily ready before the PCB layout is completed. Somewhere in the PCB layout design phase, the ASIC(s) design and manufacturing may also be ready. Now, whenever the PCB prototype is ready (t_3), the testing of it can start immediately because the test patterns are available. The same applies to the first production run (t_4), meaning that the PCB deliveries can start straight away.

If no BST is available ($\overline{BST}$), the test pattern generation will only be started at t_2 when the PCB prototype is ready. Moreover, the test preparation for in-circuit and functional test is vulnerable to changes in design or layout. Practice is that, in order to avoid waste of time and money, companies postpone the test preparation until the design verification is completed. In figure 6-4, a dotted line is drawn indicating that it is not sure when the test patterns and the test equipment are ready. This may even last till after t_4, the start of the first production run.

Relating back to figure 6-3 the introduction of BST will lower the design parts of both bars A and B, because BST allows automatic generation of the description of the Boundary-Scan registers and the automatic generation of the test pattern needed to diagnose occurring faults.

The conclusion is that BST has shortened the development time. Thus BST will certainly improve the development cycle, even in cases where concurrent engineering is already practised. The overall result is a still shorter time to market than with concurrent engineering without BST.

Prescribe Design For Testability

In some cases, certain reluctance is encountered amongst circuit designers in applying DFT and Built-In Self Tests (BISTs) into their designs. This may be due to historical reasons and/or educational backgrounds ([44], [45] and [46]).

The reluctance from an *historical* point of view may be understood if one considers the evolution in VLSI technology. In the initial phase of the VLSI design technology, the emphasis was put on area yield and performance specifications. Minimization of the circuitry per logic function was the main objective. CAD workstations provided the designers a tool for high-level synthesis of logic circuitry. But at the same time, due to economic or market pressure, management required reduced design time intervals which, as a consequence, distracted the designers from even *thinking* about DFT and BIST. The high-level synthesis development techniques were mainly performed by engineers with little or no background in DFT and BIST. In the course of time, however, DFT techniques were introduced onto CAD tools and helped the engineers to optimize the VLSI designs with respect to DFT. In the next phase of the VLSI designs built-in at-speed tests were required and additional silicon area was reserved for BISTs, including test pattern generation and compacting output responses for signature analysis (see chapter 3). Yet the results were disputed with respect to surface overhead, performance limitations and difficulty of implementation. These discussions again discouraged the VLSI designers to the extent that a certain level of knowledge of DFT and BIST implementation was required. As can be seen now, the generation of Boundary-Scan Cells around a core logic is automated to a great extent, even to the degree that the electronic circuit designer does not need specific or detailed knowledge of the various cell designs. This has accelerated the introduction of DFT and BIST techniques.

Moreover, this illustrates a better usage of capital investments. Traditionally, CAD/CAE workstations were bought for design and engineering work. Separately and independently of each other, manufacturing testers were bought. These investments can account for millions of dollars. So why not use the CAD/CAE workstations for lowering the test costs in manufacturing? This objective has become true with Boundary-Scan Test: by applying this DFT technique much less costs for testers are incurred.

These benefits can only be obtained with an *integral approach* to a new product development. The individual designers can not overview the whole cost impacts of DFT. So it may happen that one designer may like Boundary-Scan testing and applies it in the design. Another designer doesn't see DFT as his responsibility and is not likely to apply BST in the designs, *unless management makes it a mandatory design rule.*

As to the *educational* background of the electronic engineer, the subject of DFT or concurrent engineering may not yet have been taught at the institute or the university. Indeed, some curriculums are already so crowded with basic subjects that introducing a new test technique may cause another subject to be removed from the course. Sometimes, while teaching electronic design, test is regarded as a 'déjà vu' item, which is already explored and is uninteresting because everything has been solved years ago. Such an attitude does not encourage an interest in testing. The role of testing in the factory may also be easily underestimated. A designer may think of testing as an unwanted burden, which must be handled because he is the only one who understands that particular design. When a separate test department is present in a design environment, the designer may consider the test engineer as the one whose job it is to demonstrate the limits and inabilities of the designed product. Finally, in the manufacturing environment, testing is considered as just another operation which must be performed and which consumes time and resources making the product more expensive. Engineers in the factory are also not trained to cope with the testing process as a whole. But here again, the Boundary-Scan Test technology has reduced the cost of design and manufacturing resources (time, equipment) drastically, making DFT more favourable. With respect to the introduction of BST, the management, once aware of the unique opportunities of this technique, should make budgets available in order to

• learn about the new technologies,

• skill the organization in Boundary-Scan Test and

• acquire tools that support BST.

Check Impact on Reliability

Producing electronic products offers quite some challenges. Electronics manufacturers are urged to make new products with higher quality at lower costs in shorter time spans. Moreover, the management is concerned about governmental requirements on reliability issues, not only for the product design but also for the manufactured products. Companies have to meet these requirements and have to provide evidence to prove the safety of their designs. For these reasons many companies are striving for ISO 9000 certification. It is clear why DFT for printed circuit boards is playing an important role in quality and why it is imperative to meet high quality and reliability objectives.

For example, the IC/VLSI manufacturing process faces the problem of so many defects in the parts that you are lucky if half of the products are fault free [47]. Nevertheless, the required quality levels for products shipped to the customer require failure rates of a few hundred to a few thousand parts per million (PPM). Mean Time Between Failures (MTBF) for PCBs and systems are also defined. An end user experiences a fault as a fault, meaning that he does not care about the cause or reason of the fault. A particular job can not be performed because a piece of equipment is malfunctioning. Disregarding considerations of abuse or mishandling of products, the faults which can occur during the whole product life cycle have to be addressed in detail in order to take the right measures to prevent them.

A crucial factor for the product quality and reliability is the fault coverage of the various tests that weed-out a spectrum of faults. Distinction can be made between three fault types:

1. *Design Faults*
   The product is not performing the function for which it was designed. It is hard to model and to simulate all the possible design errors. No automatic test pattern generation can be done and no fault coverage can be given. The knowledge is' in the head of the designer. Design verification and/or test equipment helps to acquire the (test) data for an adequate diagnosis, but the designer has to judge the results of a test. High observability and controllability of a design increases the quality of the designer's judgement. DFT techniques (BST) help to achieve this.

2. *Manufacturing Faults*
   Manufacturing faults can be modelled to a high degree of completion. Examples are opens, shorts, component assembly faults and component defects. The test patterns can easily be generated automatically, provided that physical test access to the unit under test can be acquired. With Boundary-Scan testing all manufacturing faults can be detected and eliminated, which results in a very high delivery quality.

3. *Reliability / Life Time Faults*
   The product may cease to function due to physical or human causes, such as mishandling. A combination of test patterns for component and manufacturing tests should be able to detect these type of faults also. Reliability is further increased with the BST supported power-on self-test as described in chapter 3.

Even during system operation the reliability can be monitored, utilizing the SAMPLE mode of IEEE Std 1149.1.

Note that fault types 1 and 2 may show up in the field as a fault of type 3, due to inadequate fault coverage in design and/or manufacturing.

BST has a further positive impact on the product reliability: both the unit under test (UUT) and the tester apply the IEEE Std 1149.1. In other words, no discrepancy exists between tester and UUT, as is the case with traditional testers. Two examples are given here.

1. With in-circuit test (ICT) the overdrive technique is used to achieve the required input conditions for particular tests. This technology implies an increased chance of damaging the ICs, especially the VLSIs where long test sequences are common practice. In fact, during overdriving the ICs are used in a mode which is not specified by the manufacturer.

2. In chapter 1 it is shown that ICT with bed-of-nails fixtures no longer copes with the ongoing miniaturization on PCBs. This mismatch in technology may appear as a reliability problem as follows. The operators of the in-circuit testers know exactly which tester nails are suspected in case of a fault detection. For example a fault can be caused by a bad nail contact due to residues of soldering flux etc. As a result the operator starts interpreting the test results, instead of relying on the test equipment. In 90% of all cases the operator has right, meaning that 10% of the faults slip through to the next manufacturing phase. And such faults may even slip into the delivered products, causing perceived reliability problems in the field. With BST these problems are non-existing.

Care for Quality

What can be done to make the high quality targets obtainable? The only answer is DFT: *Design For Testability*. Given this starting point, it is unacceptable that any designer is reluctant to apply DFT, as described in a previous section. It is of prime importance to understand the quality impact of all aspects of testing, including design for testability and fault coverage targets. Testing alone does *NOT* add value to the product as seen from the customer's point of view. But if DFT is applied, then inherently the chance of failure is diminished and just that and that alone does add value (read quality) to the product. So testing must be advocated or even be made mandatory by the management as an integral part of a design/manufacturing activity and should be considered in the light of *customer satisfaction*. A customer wants quality for his money, the same money which is eventually used to pay the salaries of a company's employees!

The following points are a guide to managing company wide quality control.

› *Provide a Quality Performance Indicator*

In order to quantify the effectiveness of a test strategy one should have a kind of ruler against which the design/production results can be measured. A frequently heard slogan in respect of quality is: 'You cannot manage what you cannot measure'. For instance the targeted rate of shorts or stuck-at faults can be a quality goal. Another ruler is the price. A 30% yield in an IC production process means that the product costs about three times as much as it should.

› *Specify the Test Goals*

It is evident that the quality increases with increasing fault coverage. But the resulting quality from a test set with a certain fault coverage may prove to be dependent on the *type* of test being performed. For example, it is possible to meet some error rate but still not meet the overall quality level. Therefore all aspects of the test procedure should be understood completely before any quality targets are set. In an integral test approach the management should make it imperative that *test objectives are planned and settled even before product development is initiated*. The test plan should cover the whole product life cycle: prototype debugging, manufacturing tests and field service repairs. Boundary-Scan technology can be applied as a means for DFT.

› *Look After Education*

The understanding of producing high quality level products comes through education. A course in testing should address quality awareness. Every electronic engineer must be aware of the use of statistics or testing in process control. Overall quality control through (BST) testing should be part of every engineering course, whether the education starts at school or in the factory. This is particularly true in the factory, where a company's personnel is still its greatest asset. As a result of education very short feedback times to *prevent* faults are possible.

› *Introduce Boundary-Scan Test*

It is not the BST technology as such that results in considerable savings in product manufacturing, but the inherent spin-offs that contribute to the rise in quality. For example, one of the most eye catching properties of BST is the reuse of test software. The same test patterns generated in the design phase are also used during production tests and later in the field service. The reuse of test patterns assures a 100% consistency in testing in each life cycle phase. Another advantage is that no products are piled or sent back into the field because 'the other' test equipment could not reproduce or re-identify the fault(s). Such a procedure adversely affects the total quality. Finally, the BST test patterns generated during the design can be used during the product's whole life cycle because:

a: any encountered error will be solved immediately by the designer, b: they are for the greater part generated automatically and

c: they are designed for a fault coverage of 100%.

From the quality point of view a very important consequence of introducing BST is of course that the design engineer is "automatically" *forced* to think of the quality during the product's whole life cycle. For example if at the customer's site a fault is detected on a printed circuit board by means of a functional test, then the Boundary-Scan Test procedure should easily and automatically indicate the faulty component on that board within minutes of time. That is exactly the quality a typical customer wants: the shortest possible down-time (preferably zero) for his investments.

• *Improve the Process of the Product's Creation*
At the University of Tokyo it is taught that quality management has moved from final inspection through manufacturing control back to design improvement [48]. A product rejection in the final inspection stage does raise the delivered quality, but a high rejection rate reflects a poor design phase. So a good design offers the greatest probability for good quality output. Looked at it in this way, the creation of a product may be considered as a *process*, because right from the beginning measures are taken to keep the product well alive in the end user's environment. Therefore, it can be stated that with a great probability high quality products are produced whenever the creation process is of high quality. No need to say that BST certainly enhances the creation process since it implies design for testability.

• *Promote Concurrent Engineering*
In the previous sections it has already been shown that DFT is a prerequisite to obtain a certain quality level, but concurrent engineering is also a key factor in quality improvement, besides the time-to-market benefits. To that goal, the design and test engineers should be trained to work in teams and be able to cope with each other's problems, as well as with the problems met later on in the manufacturing process. More importantly, the Boundary-Scan Test design methodology and the related automatic test pattern generation forces the designer to cope with testing in the field, because the generated test software is reused for repairs purposes at the customer's site. It is obvious that a good test-plan forming part of the product specification best fits the methods of concurrent engineering. To enhance the quality attitude of the designers, the management has to set some constraints. First of all the designers should adhere to an integral design approach from ICs to PCBs up to systems. The designers should be supported by CAD/CAE and test tools that make the implementation transparent to technology boundaries [silicon (ICs), epoxy (PCBs), iron (racks)] and make the designs fault free, modular and reusable. This process can be helped by adhering to world wide accepted design and test standards, like VHDL, EDIF, IEEE Std 1149.1 and others.

BENEFITS OF BOUNDARY-SCAN TEST

In the previous section the role of the management is discussed in relation to product quality/reliability. This section will illustrate the benefits of applying Boundary-Scan Test. Some of the benefits are imponderable and thus hard to quantify, where others are more measurable but depend on the application. If the benefits can be made visible to the management, it is hoped that they can contribute to the introduction of BST in an organization. The most suitable time to introduce BST is at the start of a project for a new product development.

Shorter Times To Market

The impact of time-to-market has been most recognized after the publications of the Mackinsey study and other studies on this subject [49]. These publications reveal that on-average there is a reduction of 33% in the after-tax profit when products are shipped six months late as compared to a 3.5% reduction of the same profit when the product development expenses are overspent by 50%. Also, the faster a product is introduced into a competitive market, the larger potential life time it will have and hence the greater its return on investments is. Boundary-Scan improves in various ways the time-to-market of a product.

- *Concurrent Engineering*
  The role of concurrent engineering has been discussed amply in the previous section of this chapter. Figure 6-3 shows the cost bars for conventional development and concurrent engineering development; and figure 6-4 demonstrates that a further reduction in development time is even possible if BST is applied. It can be noticed that BST inherently supports concurrent engineering because the designer is concerned with the whole product life cycle. It has been concluded that concurrent engineering shortens the development times and that BST does so even further, through which the shortest possible time to market is obtained.

- *Reduced Time for Prototype Debugging*
  The types of faults the designer is confronted with are design faults (obviously), but also manufacturing type of faults like shorts and opens. However, in supporting the designer, the investments in specialized or dedicated test equipment (see a previous section) are usually kept as low as possible because when the design is ready, part of the equipment becomes obsolete and remains unused. Moreover, the designer can only do functional tests which are very ineffective due to bad fault coverage.

  The solution is BST. As mentioned before (see figure 6-4) the test programs for manufacturing faults can be easily ready in time to help the designer to debug the PCB prototype. This is especially beneficial for the design of large systems where considerable amounts of prototypes are required for the further development of the system software or the hardware/software integration. Thus reduction of this path contributes greatly in meeting the time-to-market objectives.

* *Reduced Production Ramp-up*
It frequently occurs that a commercial plans can not be kept due to start-up problems in the factory. The assistance of the designer is then requested, which is inefficient in at least two ways:

1. the dedicated test equipment of the designer is not suited to support a production line efficiently and

2. the designer does not like to do work for which he is not trained and certainly not if that work is well below his skills.

Moreover, such a working method does not guarantee high quality and it leads to overspent budgets. Here again, figure 6-4 demonstrates the value of applying BST because the availability of BST test patterns before prototyping starts ensures that prototype production and test preparation are completed well within time.

Example of Time To Market for Scan-Designed IC

Cost models have been developed to found the introduction of scan path designs. To demonstrate the impact of DFT on an electronic design, in this case an IC design, the results of one such cost model is given here [50]. As the scan design in ICs has the same impact on test costs as BST, this example may apply also to a Boundary-Scan design case. Despite the overhead in design and silicon area, the usage of scan path design is justified by the ease and speed in which high coverage test patterns can be generated, due to the circuits' combinational nature. It is understood that good test pattern generator programs are available. Thus it is possible to ship an IC product that incorporates scan test paths *before* other products that don't. It is here, in the highly competitive IC market where the utmost importance of the time-to-market is a question of "to be or to be", that a good cost/market model for IC production is a must. To show the effects of development time, a hypothetical chip is made for which the following marketing expectations are valid.

Parameter	:	Value
Gates	:	20,000
Market growth	:	0.35%/week
Price erosion	:	0.24%/week
Initial price	:	$145
Initial volume	:	1000 units/week
Product life	:	5 years

With these figures the difference in life time profit was calculated as function of the slippage in first delivery in weeks. The outcome is shown in figure 6-5.

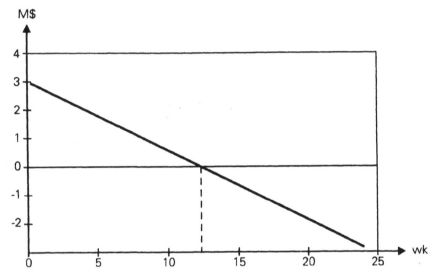

Fig. 6-5   Difference in profit due to design delay

If the design and testing of a regular product slips about three moths, then a loss of profit of $3 Million over the five years life time may be expected. Or put it the other way round, the chip area savings of a non-scan design are wasted if this results in a three month longer test preparation time, which comes to about $230,000 per week.

Yes, this is just a model. But once more it underlines the enormous advantage of applying DFT techniques in relation to time to market. Evidently, the same applies for BST.

Lower Capital Investments

In the previous section it has become clear that an introduction of Boundary-Scan Test technology also implies the application of DFT and concurrent engineering. As a consequence the efforts in the design phase have to be increased. Computer power for the software developers also requires considerable investment (see previous sections), in addition to the manpower costs. In order to obtain the highest added value from the designers, they should be equipped with CAD/CAE workstations and supporting software packages.

These computer costs may easily be millions of Dollars, but are nevertheless needed for design activities. The point is to exploit these tools to their full extent and not just for design project. These tools should support both DFT for ICs and/or PCBs designs (preferably automated) and test preparation activities for which the automatic inputs of the CAD/CAE stations are very important in order to work fault free. For instance the needed files for EDIF netlists as used for the technical product documentation can easily be transformed in BSDL data sheets. This will also avoid the tedious manual data entries, thus assuring the quality of data transfer.

The introduction of BST will further reduce the investments for testing, particularly in the manufacturing phase of the product life cycle. This reduction has two causes. Firstly, the price of the BST tester is much lower than that of a traditional ICT tester. The prices come down from hundreds to tens of thousands of Dollars. One of the reasons for this is the number of needed test pins (nails): a multiple of 5 (e.g. 25) is used in the BST case and thousands in the ICT case. Moreover, BST testers are high SW-intensive, which make them liable to price erosions. Secondly, for BST less testers are needed due to shorter fault diagnosis times, through which the factory throughput per tester is increased.

Cost Considerations for ICs

The extra costs for implementing Boundary-Scan Test provisions in ICs plays an important role. For ASIC designs a preliminary estimate can be made. Suppose implementing the BST logic requires 15 gates per pin plus a fixed 500 gates for the TAP Controller. If the core logic of the ASIC contains k gates and the package has n pins, the relative extra costs can be expressed as a percentage p of overhead silicon area:

$$p = \frac{500 + 15 \cdot n}{k} \times 100\%$$

Investigation of some examples yields the following:

for an ASIC which has 68 pins and 10,000 gates: $p = 15.2\%$
for an ASIC which has 208 pins and 200K gates: $p = 1.81\%$

Some additional effects which can be taken into account may influence the estimation of the real overhead. For instance the BS cells can be integrated in the standard I/O cells giving a drastic reduction in used silicon. If not all the usable gates are consumed, then the overhead is actually not adding costs unless the total number is going over a limit and the next part number in a range has to be used.

For standard VLSIs, the Boundary-Scan implementation may incur an overhead of just 0.3%. This percentage has been published already in 1990 by Motorola for its 68040 chip [51]. The 68040 possesses 174 pins and 1.2M transistors.

For the standard glue logic (buffers, octals) the overhead is much higher and consequently their prices may be multiplied when compared to the same functionality *without* BST. But on board level, where just a few of these functions are used, the prices of the applied VLSIs and ASICs are an order of magnitude more expensive, so in total the relatively costly BST supporting buffers or octals have a minor effect on the PCB price.

In conclusion, the extra costs added to ICs vary depending on complexity, number of pins, etc. but are not of such a size to restrict the introduction of Boundary-Scan Test. Moreover, standard VLSIs tend to contain BST provisions anyhow and can not be bought without Boundary-Scan provisions.

PCB Cost Considerations

To demonstrate the cost effectiveness of the introduction of Boundary-Scan testing a comparison is made based on the characteristics of a professional product division of Philips Electronics. It concerns the production of printed circuit boards (PCBs). The comparison includes the design of the Boundary-Scan Registers in the applied ASICs, the generation of the test programs, the testing/diagnosis time and the repairs of the defective PCBs, the latter including the tester investments and cost of ownership (amortizing, insurance, power consumption, maintenance). Further it is assumed that one lot of PCBs comprises full Boundary-Scan boards whereas the others have no BST applied.

Table 6-2 gives the overview. In the first row the costs for the Boundary-Scan Registers in each ASIC is calculated. Depending on the complexity and the number of I/O pins (see former subsection), for this application 1% extra costs on an ASIC was the real figure.

The reduced test programming time for the Boundary-Scan models with respect to the other models (1:6) results in drastically lower costs and shows again the big advantage of the BS technology: along with the description of the Boundary-Scan cells comes the automatic test pattern generation. Even the 50 hours in this example is considered to be a conservative estimate.

Table 6-2   Comparative test costs model

	Boundary-Scan	Non Boundary-Scan
Implementation costs BSR 200k boards per year, 12 ASICs per board, 1% of $25 per ASIC $\Rightarrow$	**$600k**	**None**
Test program generation 30 PCB types/year @ $50 per hour	50 hours/type **$75k**	300 hours/type **$450k**
Diagnostics 70% yield of 200k boards: $\Rightarrow$ 60k boards to be repaired @ $25 per hour	2 min. per repair: $\Rightarrow$ 2000 hours **$50k**	10 min. per repair: $\Rightarrow$ 10,000 hours **$250k**
Number of testers 200k boards/year, test time 3 min./board: $\Rightarrow$ 10,000 hours 3 shifts per tester yields 5000 hours/year/tester $\Rightarrow$	Plus diagnosis time and retest: $\Rightarrow$ 15,000 hours **3 testers**	Plus diagnosis time and retest: $\Rightarrow$ 23,000 hours **5 testers**
Tester costs Investment $\Rightarrow$ Cost of ownership 33% per year $\Rightarrow$	$75k **$75k**	$500k **$830k**
Fixtures costs 30 fixtures:	@ $5k $\Rightarrow$ **$150k**	@ $15k $\Rightarrow$ **$450k**
Labour costs @ $25 per hour $\Rightarrow$	15,000 hours **$375k**	23,000 hours **$575k**
YEARLY TOTAL	**$1325k**	**$2555k**

For other test preparation methods (in-circuit or functional test) the number of hours is much higher in reality and tends to increase with the complexity of the boards. The diagnostics of the design with BST showed to be much faster than for functional test, especially when the BST includes the on-board functional test of the ASICs (INTEST, RUNBIST). Therefore, for the *digital* part of the unit under test (UUT), the Boundary-Scan tester is assumed to be the *structural* tester (i.e. testing for manufacturing faults). The tester for testing the *analog* part of the UUT is a Manufacturing Defects Analyzer (MDA). Such a combination of testers costs a fraction of a functional in-circuit tester. Fixtures for MDA concern first of all less pins and secondly no restrictions in wiring lengths and type of cables (coax). This results in a lower fixture price. The rows for the tester and fixture costs in the table show very clear why BST was introduced in the first place: the test costs with respect to the non-BST model are (in this example) reduced by 80%! Here the four TAP wires for TCK, TMS, TDI and TDO replace many bed-of-nails probes and their electronic signalling.

Again, this is a modelled example, but stating that the introduction of Boundary-Scan testing saves about half of the factory investments for testing is certainly not exaggerated and also the comparisons listed per row in table 6-2 are not too far from reality. This table may serve as a guide for those readers who want to examine these figures in their own environment. In addition to these calculations there are the spin-offs of Boundary-Scan Test, like a shorter time to market, higher quality level, etc. These benefits can easily override those from the calculations in this table.

A similar result of a BST test analysis is reported by Alcatel Bell [52]. In this paper it is concluded that implementing BST increases the price of ASICs with 5%. In the *engineering* phase the board design time is 'slightly' increased, but the test verification and test preparation costs are reduced with 20% and 50% respectively. The introduction of BST in the *product test* phase of the product (PCB) further reduced the test costs for the same two reasons as mentioned above.

1. Reduction of test equipment costs because less equipment is needed of which the price is much less, and

2. The diagnosis time for BST is only one third as compared to in-circuit testing.

Therefore it can be concluded that, depending on the board design, an overall cost reduction of 18% – 35% is obtainable by introducing BST in the designs.

SUMMARY

This chapter highlights items that should have the attention of the management of electronics companies. The introduction of Boundary-Scan Test influences the management's role in so much that giving some attention to new working methods will lead to an improved factory profitability and return on investments. Various companies in this field have proven this statement, which can be concluded from the introduction of significantly lower priced electronic products with increased functionality and quality, giving these companies a lead in the market.

The next points summarize the conclusions drawn in this chapter.

1. *Cheaper products*

   - BST causes lower capital investments for test equipment.

   - BST allows shorter test times due to a fault coverage of nearly 100% and strong diagnostics software.

2. *Time-to-market is shorter*

   - Shorter development times because BST implies DFT, encourages concurrent engineering and speeds up prototype debugging.

   - Faster production ramp-up because the test programs are ready in advance.

3. *Improved product quality and reliability*

   - The products and the test equipment use the same BST technology, which is in itself technology independent.

   - Lower product down time (MTBF) due to improved manufacturing methods; no ad-hoc approach at delivery time.

   - Quicker repair at the customer's site due to a very high fault coverage of the fault-free BST test software and diagnostic reporting.

APPENDIX

This appendix contains, in alphabetical order, a concise representation of the specifications of the BST architecture, the test access port and controller, the instruction register, the test data registers, the instructions and the documentation requirements, all according to the IEEE Std 1149.1. The principles and the operation of these subjects are described in chapter 2. The appendix is for those readers who do not have the formal specifications of the IEEE Std 1149.1-1990 [1] to hand.

BOUNDARY-SCAN REGISTER

According to the IEEE 1149.1 standard the following provisions for the Boundary-Scan Register are mandatory.

- A Boundary-Scan Register cell must be connected between each digital system pin and the on-chip logic.

- The applied Boundary-Scan cell must be able to observe and control the state of the signal at the component pin and, if applicable, also to control the on-chip system logic.

- A (combination of) Boundary-Scan cell(s) must be able to observe and control signals at unidirectional and bidirectional system pins, at 2-state and open collector pins and at 3-state system output pins. Where appropriate, the same applies to the chip system logic.

- The above provisions also apply to the digital connections between an analogue–digital interface and the on-chip system logic of an IC containing analog circuitry.

- If, in a particular programmed system configuration a Boundary-Scan cell is unable to meet the above rules, then it must be designed such that:
  a) the results of the *EXTEST, INTEST* and *RUNBIST* instructions are not affected by the data shifted into the cell, and
  b) the data value shifted out of the cell following the *Capture-DR* state of the TAP Controller is either the value previously shifted in or a constant (0 or 1).

- For programmable components, the length of the Boundary-Scan Register must be independent of the way that the component is programmed.

- One Boundary-Scan cell may be used to control output buffers at more than one system pin in cases where one output signal from the on-chip logic enables or controls the logic output at those system pins.

- One Boundary-Scan cell may be used for both the input and output pins in cases where a system input pin is *solely* used to control data at a system output pin, for example to provide an output enable or direction control signal for a 3-state or bidirectional pin.

Cells at 2-State Output Pin

The cell specifications for the 2-state output pins are the following.

- Each cell must contain a shift-register stage with a latched parallel output.

- As a response to the *SAMPLE/PRELOAD* or the *INTEST* instruction the shift-register stage loads the data output from the on-chip logic on the rising edge (positive slope) of the TCK pulse when the TAP Controller is in its *Capture-DR* stage.

- A sample signal must not be inverted, i.e. if the on-chip logic supplies a logic 0 signal, a logic 0 must also be shifted towards the TDO.

- Data generated by the on-chip logic is supplied to the output pin without modification by the cell, for example when the Boundary-Scan Register is not selected or after a *SAMPLE/PRELOAD* instruction.

- The cell may be designed to act as part of a signal analyzer for test results from the on-chip system logic when the *RUNBIST* instruction is selected.

- When the *RUNBIST* instruction is selected, the cell must not cause any change in the signal driven to the output pin.

- As a response to the *EXTEST* or *INTEST* instruction, the signal driven to the system output pin must change at the falling edge (negative slope) of the TCK pulse when the TAP Controller is in its *Update-DR* state.

- The signals previously shifted into the cell during TDI must not change when they are driven to the output pin unchanged (a logic 0 remains a logic 0). The applicable instructions are *EXTEST*, *INTEST* or *RUNBIST*.

Cells at 3-State Output Pin

The cell specifications for the 3-state output pins are the following.

- The cell that controls the active/inactive drive may be designed to select the inactive drive when the TAP controller is in its *Test-Logic-Reset* state.

- Two cells must control this pin: one to supply the data and one to select the active or inactive drive state.

- Both cells must meet the specifications for cells at the 2-State Output pins, with the exception of the last two specifications, which are mandatory only for the *EXTEST* instruction.

- When the *INTEST* or *RUNBIST* instructions are selected, then either one of the two following requirements must be met.
  1) The output pin is placed in the inactive drive state (e.g. high impedance).
  2) The last three specifications listed for the 2-State Output Pin must be met, if and when applicable.

- The cell may be designed to act as part of a signal analyzer for test results from the on-chip system logic when the *RUNBIST* instruction is selected.

Cells at Bidirectional Pin

The cell specifications for a bi-directional pin are the following.

- Cells are provided to select between input and output of signals, and which observe or control the data value at the pin.

- For 3-state bidirectional pins the input/output control cell must meet the requirements for the cell that drives the state of 3-state output pins.

- The input Boundary-Scan Cell must meet the same requirements as the input cell (see above).

- In output operation, the output cells must meet the requirements of the equivalent 2-State Output Pins, or where applicable the 3-State Output Pins.

Cells at System Clock Input Pin

- Each cell must contain a shift-register stage.

- As a response to the *SAMPLE/PRELOAD* or the *EXTEST* instructions, the shift-register stage loads the data presented at the pin on the rising edge (positive slope) of the TCK pulse when the TAP controller is in its *Capture-DR* state.

- Data presented at the input pin is supplied to the on-chip logic without modification by the cell, for example when the Boundary-Scan Register is not selected or after a *SAMPLE/PRELOAD* instruction.

- When the *RUNBIST* or *INTEST* instruction is valid, the clock signal applied to the on-chip logic must be one of the following.
  1) The signal received at a system clock input pin.
  2) The TCK signal, such that the on-chip logic can only change when the TAP Controller is in its *Run-Test/Idle* state.
  3) For the *INTEST* instruction only: a signal that is supplied by shifting the Boundary-Scan Register.

- The signal applied to the on-chip logic as a response to the *EXTEST* instruction is forced to a high or low logic level independently of the data present in the shift-register stage or at the system clock input pin.

Cells at System Input Pins

- Each cell must contain a shift-register stage.

- As a response to the *SAMPLE/PRELOAD* or the *EXTEST* instructions, the shift-register stage loads the data presented at the pin on the rising edge (positive slope) of the TCK pulse when the TAP controller is in its *Capture-DR* state.

- A sample signal must not be inverted, i.e. a logic 0 applied to the input pin causes (later) a logic 0 to be shifted towards the TDO.

- Data presented at the input pin is supplied to the on-chip logic without modification by the cell, for example when the Boundary-Scan Register is not selected or after a *SAMPLE/PRELOAD* instruction.

- The signal driven to the on-chip system logic as a response to an *INTEST* instruction must be the same as that previously shifted into the BSC and must have the same result as if it were supplied directly to the on-chip system input during system operation.

- The cell may be designed such that the signal applied to the on-chip logic as a response to the *EXTEST* or *RUNBIST* instruction is the same as if it were obtained from the shift-register after a *INTEST* instruction.

- The cell may be designed such that the signal applied to the on-chip logic as a response to the *EXTEST* instruction is forced to a high or low logic level independently of the data present in the shift-register stage.

- The design of the cell must cause no interference with the operation of the on-chip self-test, when this is selected by the *RUNBIST* instruction.

- The cell may be designed to act as a generator of test patterns for the on-chip system logic when the *RUNBIST* instruction (or an alternative self-test instruction) is selected.

- The cell may be controlled such that, during the execution of a self-test instruction other than the *RUNBIST*, data may flow back and forth between the system pins and the on-chip system logic without modification.

DEVICE ID REGISTER

The specifications for the Device ID Register (if present) are the following.

- The Device ID register must be a shift-register based path that has a parallel input but no parallel output.

- The Device ID Register is set at the rising edge (positive slope) of the TCK pulse when the TAP Controller is in its *Capture-DR* state. The same code is shifted out towards the TDO at a subsequent shifting process.

- The vendor-defined identification must contain four fields, which can be read using the *IDCODE* instruction.

- For user-programmable components, a supplementary user-programmable identification code data must be provided that can be loaded into the device identification register using the *USERCODE* instruction.

DOCUMENTATION REQUIREMENTS

Boundary-Scan Register

The following information is required in addition to that listed under the Test Data Registers (above).

- The correspondence between boundary-scan register bits and system pins, system direction controls, or system output enables.

- Whether each pin is an input, a 2-state output, a 3-state output, or a bidirectional pin.

- For each boundary-scan register cell at an input pin, whether the cell can apply tests to the on-chip system logic.

- For each boundary-scan register cell associated with an output or direction control signal, a list of the pins controlled by the cell and the value that shall be loaded into the cell to place the driver at each pin in an inactive state or will be observed using the *SAMPLE/PRELOAD* or *INTEST* instructions when the on-ship system logic causes the driver to be inactive.

- The method by which single-step operation is to be achieved while the *INTEST* instruction is selected, if this instruction is supported.

- The method of providing clocks to the on-chip system logic while the *RUNBIST* instruction is selected, if this instruction is supported.

Device-Identification Register

Where a device identification register is included in a component, the following information is required in addition to that listed under the Test Data registers (above).

- The value of the manufacturer's identification code.

- The value of the part number code.

- The value of the version code.

- The method of programming the value of the supplementary identification code, where required.

Instruction Register

The following information pertaining to the instruction register is required.

- Its length.

- Whether instructions have a parity bit and, if so, its location.

- The pattern of fixed values loaded into the register during the *Capture-IR* controller state.

- The significance of each design-specific data bit presented at a parallel input, where provided.

Instructions

For each public instruction offered by a component, the following information is required.

• The binary code(s) for the instruction.

• A list of test data registers placed in a test mode of operation by the instruction.

• The name of the serial test data register path enabled to shift data by the instruction.

• A definition of any data values that shall be written into test data registers prior to selection of the instruction, and the order in which these values shall be loaded.

• The effect of the instruction. Any system pins whose drivers become inactive as a result of loading the instruction should be clearly identified.

• A definition of the test data registers that will hold the result of applying a test and of how they are to be examined.

• A description of the method of performing the test and of how data inputs and their corresponding data outputs are to be computed.

Performance

The performance of the test logic should be fully defined, including the following information:

• The maximum acceptable TCK clock frequency.

• A fully set of timing parameters for the test logic.

• The logic switching thresholds for TAP input and output pins.

• The load presented by the TCK, TMS, TDI, and TRST* pins.

• The drive capability of the TDO output pin.

• The extent to which the TDO driver may be overdriven when active (for example while using an in-circuit test system).

• Whether TCK may be stopped in the logic 1 state.

Self-Test Operation

For each instruction that causes operation of a self-test function, the following information is required.

• The minimum duration (for example a number of TCK cycles) required to ensure completion of the test.

• A definition of the test data registers whose states are altered during execution of the test.

• A definition of the results of executing the self-test on a fault-free component.

• An estimate of the percentage (for example to the nearest 5%) of the single stuck-at faults in the component's circuitry that will be detected *or* a description of the functional operation and the circuitry exercised.

Test Data Register

For each test data register available for public use and access in a component, the following information is required.

• The name of the register, used for reference in other parts of the data sheet.

• The purpose of the register.

• The length.

• A full description of the operating modes of the register.

• The result of setting each bit at the parallel output of the register.

• The significance of each bit loaded from the parallel input of the register.

Timing Definitions

Figures A-1 and A-2 give an example of how set-up and holding time parameters should be measured, relative to the TCK and the reference voltage V_{ref}. These timing parameters are required for the TMS, TDI and TDO signals. They have also to be specified for system pins that can be driven from the test logic.

An example is the system data input set-up time T_{setup} for the Boundary-Scan register before the rising edge of the TCK while the TAP Controller is in its *Capture-DR state* (see figure A-1).

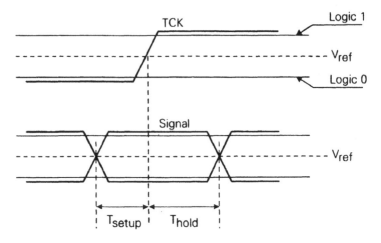

Fig. A-1   Measuring Set-up and Hold times

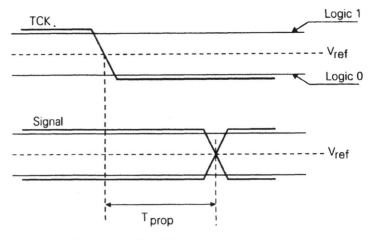

Fig. A-2   Measuring propagation delays

INSTRUCTION REGISTER

The rules prescribed by the IEEE Std 1149.1 for the Instruction Register (IR) are the following.

- The behaviour of the IR in each TAP Controller state must be as listed in chapter 2, Table 2-1, along with the description of the Instruction Register.

- When a new instruction is loaded into the parallel output of the IR, the actions resulting from of the preceding instruction cease. This loading into the parallel latches must occur at the falling edge (negative slope) of TCK in the *Update-IR* controller state.

- When the TAP Controller comes in the *Test-Logic-Reset* state, then the *IDCODE* instruction must be latched into the IR on the falling edge (negative slope) of TCK. If no device ID register is available, then the *BYPASS* instruction must be selected in this case.

- If a TRST* input to the latches is provided (see chapter 2, figure 2-13), then, when a negative signal is applied, the current instruction in the latches is changed asynchronously to *IDCODE*, or *BYPASS* if no device ID register is present.

INSTRUCTIONS

BYPASS Instruction

- Each component (IC) must provide a *BYPASS* instruction code.

- The binary code for the *BYPASS* instruction must be {111...1} (all 1s), thus a logic 1 is entered into every instruction register cell. However, additional binary codes are also permitted.

- The selected Bypass Register must be connected in the serial TDI/TDO path while the TAP Controller is in its *Shift-DR* state.

- When the *BYPASS* instruction is current, all test data registers that can operate in either test or system modes must perform their system function. Also, the test logic must not influence the operation of the on-chip system logic.

CLAMP Instruction

The following rules apply when the optional *CLAMP* instruction is provided.

- The *CLAMP* must select the Bypass register to be connected between TDI and TDO in the *Shift-DR* state.

- When the *CLAMP* instruction is selected, the value of all signals at the system output pins must be completely defined by the data held in the Boundary-Scan Register.

- The states of the parallel output latches in the Boundary-Scan cells located at the system output pins (2-state, 3-state or bidirectional) must not change when the *CLAMP* instruction is selected.

- When the *CLAMP* instruction is selected, the core logic must be controlled such that it can not be damaged by signals received at the system input or clock input pins, for example by setting the core logic in its reset or hold state.

- The component designer is allowed to select binary value(s) for the *CLAMP* instruction.

EXTEST Instruction

The following specifications must be met for the IEEE 1149.1 standard:

- The component must provide the *EXTEST* instruction.

- The binary code for the *EXTEST* instruction must be {000...0} (all 0s), thus a logic 0 is entered into every instruction register cell. However, additional binary codes are also permitted.

- The *EXTEST* instruction selects only the Boundary-Scan Register required for access via the serial TDI/TDO path, i.e. no other test data register may be connected in series with the Boundary-Scan register. The instruction occurs when the TAP Controller is in its *Shift-DR* state.

- While the *EXTEST* instruction is executed, the on-chip logic must be protected against any damage caused by the signals received at the system input or clock pins.

- Signals arriving at the component's input pins are loaded in the Boundary-Scan Register at the rising edge (positive slope) of the TCK pulse while the TAP Controller is in its *Capture-DR* state.

- When the *EXTEST* instruction is selected while the TAP Controller is in its *Capture-DR* state, the data loaded into the Boundary-Scan register cell at the output side should be independent of the operation of the on-chip logic.

- Test signals which are sent from the Boundary-Scan Registers through the output pins can only change on the falling edge (negative slope) of the TCK pulse while the TAP Controller is in its *Update-DR* state. These test signals must be completely defined by the test data previously shifted into the Boundary-Scan Register.

HIGHZ Instruction

The following rules apply when the optional *HIGHZ* instruction is provided.

- The *HIGHZ* must select the Bypass register to be connected between TDI and TDO in the *Shift-DR* state.

- When the *HIGHZ* instruction is selected, all system logic outputs (including 2-state, 3-state or bidirectional pins) must immediately be placed in an inactive-drive state, for example high impedance.

- When the *HIGHZ* instruction is selected, the core logic must be controlled such that it can not be damaged by signals received at the system input or clock input pins, for example by setting the core logic in its reset or hold state.

- The component designer is allowed to select binary value(s) for the *HIGHZ* instruction.

IDCODE Instruction

If a Device ID Register is included in a component, the following specifications must be met for the IEEE 1149.1 standard.

- Manufacturers must provide an *IDCODE* instruction for components that include a Device Identification Register.

- When the *IDCODE* instruction is selected, only the Device Identification Register is connected between the serial TDI/TDO path, i.e. no other test data register may be connected in series with the Device Identification Register. The instruction occurs when the TAP Controller is in its *Shift-DR* state.

- When the *IDCODE* instruction is selected, the vendor identification code is loaded into the Device Identification Register on the rising edge (positive slope) of the TCK while the TAP Controller is in its *Capture-DR* state.

- When the *IDCODE* instruction is selected, all test data registers that can operate in either test or system modes must perform their system function. Also, the test logic must not influence the operation of the on-chip system logic.

INTEST Instruction

The following specifications must be met for the IEEE 1149.1 standard.

- The *INTEST* instruction selects only the Boundary-Scan Register required for access via the serial TDI/TDO path, i.e. no other test data register may be connected in series with the Boundary-Scan register. The instruction occurs when the TAP Controller is in its *Shift-DR* state.

- Test signals which are sent from the Boundary-Scan Registers through the output pins can only change on the falling edge (negative slope) of the TCK pulse while the TAP Controller is in its *Update-DR* state. These test signals must be completely defined by the test data previously shifted into the Boundary-Scan Register.

- All non-clock signals driven into the on-chip logic (IC) must have been shifted into the Boundary-Scan Register stages.

- Signals from the on-chip logic are loaded into the Boundary-Scan Register at the rising edge (positive slope) of the TCK pulse while the TAP Controller is in its ·*Capture-DR* state.

- The on-chip system logic must be capable of single-step (slow-speed) operation while the *INTEST* instruction is selected.

- When the *INTEST* instruction is selected while the TAP Controller is in its *Capture-DR* state, the loaded data should be independent of the operation of the off-chip circuitry or board-level interconnections.

RUNBIST Instruction

The following specifications must be met for the IEEE 1149.1 standard.

- The TAP Controller must be in its *Run-Test/Idle* state when the *RUNBIST* instruction is selected.

- No provisions need to be made for seed values to be entered in other than Boundary-Scan test data registers.

- The duration of the execution of the *RUNBIST* must be a specified item (e.g. specified as a number of system clock pulses).

- Following the specified minimum duration, the test result observed by loading and shifting the test data register selected by the *RUNBIST* must be constant.

- The data of test results shifted out of the component (IC) in response to the *RUNBIST* instruction must not be altered by the presence of faults in off-chip system logic, board-level interconnections etc.

- The state of the parallel output registers at the IC output pins (2-state, 3-state, bidirectional) must not change while the RUNBIST is selected.

- The test data register into which the results of the self-test are to be loaded must be connected for access in the serial TDI/TDO path while the TAP Controller is in its *Shift-DR* state.

- The results of the self-test must be loaded into the test data register (connected in the serial TDI/TDO path) no later than the rising edge (positive slope) of the TCK pulse while the TAP Controller is in its *Capture-DR* state.

- Use of the *RUNBIST* must give the same result in all versions of the component (IC).

Public Instructions

- All public instructions must be available to the purchasers.

- Instructions must *completely* define all test data registers that may operate and interact with the on-chip logic for as long as the instruction is current. Non-selected test data registers must be controlled so that they do not interfere with the operation of the on-chip logic, or with the operation of the test data registers which have been selected.

- Each instruction enables a single serial test data path along which data are shifted from TDI towards TDO while the TAP Controller is in its *Shift-DR* state.

- Instruction codes which are no longer used must be equivalent to the *BYPASS* instruction.

- If the optional Device ID Register is included in an IC, then the *IDCODE* instruction must be provided by the vendor. If the IC is user-programmable, the *USERCODE* must also be provided by the vendor.

- All components for which the vendor claims compliance with the IEEE 1149.1 standard must contain the following public instructions: *BYPASS*, *SAMPLE/PRELOAD* and *EXTEST*. If the optional instructions *INTEST* and/or *RUNBIST* are available, then they must also comply with the IEEE 1149.1 standard.

- The mode of operation of a test data register may be defined by a combination of the current instruction and further control information contained in test data registers.

- A component design may offer public instructions additional to the IEEE 1149.1 standard to allow the component purchaser access to design-specific instructions.

SAMPLE/PRELOAD Instruction

The following specifications must be met for the IEEE 1149.1 standard.

- The component must provide the *SAMPLE/PRELOAD* instruction.

- The *SAMPLE/PRELOAD* instruction only addresses the Boundary-Scan Registers required for access via the serial TDI/TDO path, i.e. no other test data register may be connected in series with the Boundary-Scan register. The instruction occurs when the TAP Controller is in its *Shift-DR* state.

- During the execution of the *SAMPLE/PRELOAD* instruction the on-chip-logic operation is not hampered in any way.

- When the *SAMPLE/PRELOAD* instruction is selected, the signal states at the input and output pins are loaded in the Boundary-Scan cells on the rising edge (positive slope) of the TCK pulse while the TAP Controller is in its *Capture-DR* state (*SAMPLE* action).

- When the *SAMPLE/PRELOAD* instruction is selected, the data present in the Boundary-Scan shift-register stage is loaded into the parallel output/latch at the falling edge (negative slope) of the TCK pulse while the TAP Controller is in its *Update-DR* state (*PRELOAD* action).

USERCODE Instruction

The IEEE 1149.1 standard requirements for the USERCODE are the following.

- Manufacturers must provide a *USERCODE* instruction for components that include a Device Identification Register and are user-programmable such that the programming cannot be otherwise provided to the test logic.

- When the *USERCODE* instruction is selected, only the Device Identification Register is connected between the serial TDI/TDO path, i.e. no other test data register may be connected in series with the Device Identification Register. The instruction occurs when the TAP Controller is in its *Shift-DR* state.

- When the *USERCODE* instruction is selected, the user-programmable identification code is loaded into the Device Identification Register on the rising edge (positive slope) of the TCK while the TAP Controller is in its *Capture-DR* state.

- When the *USERCODE* instruction is selected, all test data registers that can operate in either test or system modes must perform their system function. Also, the test logic must not influence the operation of the on-chip system logic.

TAP CONTROLLER

The following rules for initialization of the TAP Controller are set by the IEEE 1149.1 standard.

- The TAP Controller must be forced into its *Test-Logic-Reset* state at power-up either by the use of the TRST* signal or as a result of circuitry built into the test logic. If the TRST* is used, the initialization must occur asynchronously when the TRST* input changes to the low level. If the built-in circuitry, the result must be equivalent to that which would be achieved by using TRST*, i.e. application of a logic 0 to the TRST* input.

- The TAP Controller must not be initialized by operation of any system signal, such as a system reset.

TEST DATA REGISTERS

In this section, some of the *common* design requirements are described.

- Each test data register must have a unique name.

- In the *Shift-DR* state of the TAP Controller the data applied to the TDI is shifted towards the TDO, which must be done without inversion of the data.

- The length of any test data register must be fixed, independent of the instruction by which it is accessed. A named test data register may share registers from other system logic, but each part and each addressable combination of parts must have a unique name. The resulting design must comply fully with the IEEE 1149.1 standard.

- For programmable components, the length of each test data register must be independent of the way in which the component is programmed.

- The circuitry contained in test data registers may be used to perform system functions when test operation is not required, provided it is not in contradiction with the IEEE 1149.1 standard.

- Each instruction must identify a test data register that will be serially connected between TDI and TDO. Such a register shifts data one stage towards TDO on each rising edge (positive slope) of the TCK in the TAP Controller's *Shift-DR* state.

- An instruction may select a test operation with which more than one test data register is involved. This must then be done sequentially, step by step, because at any time there can be only one test data register between the TDI and the TDO in the *Shift-DR* controller state.

- Test data registers which are not selected for a test operation must be configured so that they do not interfere with the operation of the on-chip logic or else perform their *system* function (if one exists). The same applies when the TAP Controller is in its *Test-Logic-Reset* state.

- Data are loaded from a parallel input into the test data register in the TAP Controller's *Capture-DR* state on the rising edge (positive slope) of the TCK.

- If a test data register is provided with a latched parallel data output, the data must be latched into the parallel output buffer on the falling edge (negative slope) of the TCK. The TAP Controller is then in its *Update-DR* or *Run-Test/Idle* state, as appropriate.

- Where an 'internal' test execution (e.g. a self-test) is required, this must occur while the TAP Controller is in the *Run-Test/Idle* state.

- When in a given TAP Controller state no operation of a test data register is required by an instruction, this register must retain its last state unchanged.

GLOSSARY

The following glossary of terms and abbreviations reflects terms used in this book but may not reflect terms commonly used in electronics. Terms printed in **bold** in the descriptions below are defined themselves in the glossary.

2-State Output Pin –
> A device output pin at which the signal can be only at a logical high or a logical low level at any given instant.

3-State Output Pin –
> A **2-state output pin** which can *also* be held in an inactive state, for example at high impedance, in which the pin's output logic can not determine the logical level at the connected external circuitry.

Aliasing Test Result –
> Response of a test of a particular net where a faulty result looks the same as a fault-free result of the test on another net, so that no unique diagnosis of the fault location is possible.

ASIC   Application Specific Integrated Circuit

ATE   Automatic Test Equipment

ATPG   Automatic Test Pattern Generator

Bidirectional Pin –
> A device contact pin at which a signal can be either received from or transmitted to the connected external circuitry.

Binary Counting Sequence –
> A test method at which a net to be tested is supplied with a unique sequence of test bits, such that the applied binary sequence reflects the decimal number subsequently given to each net.

BITL   **BST** Integrated Testvector List

Blind Interrogation –
> Retrieving information out of a component (e.g. from a device's identification register) without further knowledge of the operation of the component concerned, thus providing for a quick check.

BPV   **BST** Parallel Vector

BS   Boundary-Scan

BSC   Boundary-Scan Cell

BSDL   Boundary-Scan Description Language

BSR   Boundary-Scan Register

BST   Boundary-Scan Test

BTPG™   Boundary-scan Test Pattern Generator (Trade Mark of Philips Industrial Electronics B.V.)

BTSL   Boundary-scan Test Specification Language

BYPASS   Instruction for **BST** defined in the **IEEE Std 1149.1**.

CLAMP   Optional instruction for **BST** defined in the **IEEE Std 1149.1**.

Capture   A **BST** operation at which data present on an input signal line is
                 loaded into a memory element of the **BSR**.

Cluster   In the context used in this book, a piece of logic on a **PCB**
                 surrounded by **BST** compatible components but which itself is not
                 compatible to the **IEEE Std 1149.1**.

COB   Chip On Board; chip directly mounted on printed circuit board,
                 without being packaged first.

Confounding Test Result −
                 Response of a test of a net by which the faulty test result looks the
                 same as a faulty result of the test on another net, so that no unique
                 diagnosis of the fault location is possible.

Core logic   As opposed to the **test logic**, the functional logic performing the
                 wanted non-test functions.

DFT   Design For Testability

DR   Data Register

DUT   Device Under Test

EDIF   Electronic Data Interchange Format; an international standard
                 language to describe electronic circuits.

EXTEST   External test, an instruction for **BST** defined in the **IEEE Std 1149.1**.

HIGHZ   Optional instruction for **BST** defined in the **IEEE Std 1149.1**.

ICT   In-Circuit Test

IDCODE   Identity code, an optional instruction for **BST** defined in the **IEEE
                 Std 1149.1**.

IEEE Std 1149.1 −
                 Abbreviation for 'IEEE Standard Test Access Port and Boundary-
                 Scan Architecture' [1].

Interconnect Test −
                 A test for the correctness of all **PCB** level interconnects.

INTEST   Internal test, an optional instruction for **BST** defined in the **IEEE Std
                 1149.1**.

IR   Instruction Register

JEDEC   Joint Electron Device Engineering Council, an international body
                 maintaining standards on electrical and electronic devices.

JTAG   Joint Test Action Group; group of test engineers representing world-
                 wide electronics companies preparing the **BST** method that resulted
                 in the **IEEE Std 1149.1**.

LSB   Least Significant Bit

MCM   Multi Chip Module

MDA   Manufacturing Defects Analyzer; tester for electronic devices (**PCBs**)
                 in a manufacturing environment.

MSB   Most Significant Bit

MTBF   Mean Time Between Failures

Net   Configuration of connected-through copper tracks/wires on a **PCB**.

Open	A name of a fault caused by an interrupted electrical connection, for example between a component pin and the **PCB** track or caused by a broken copper track.
PCB	Printed Circuit Board
PLB	Programmable Logic Devices
PPM	Parts Per Million
Private Instruction –	An **IEEE Std 1149.1** instruction which is solely meant for internal use by the component (IC) manufacturer.
PTH	Plated Through Hole; a **PCB** at which the drilled holes are internally metal plated.
PTV	Parallel Test Vector, a number of test signals which are supplied simultaneously to various test nodes.
Public Instruction –	An **IEEE Std 1149.1** instruction providing the end user of a component (IC) with access to test features, documented by the component supplier.
PWB	Printed Wiring Board
RUNBIST	Run Built-In Self Test, an optional instruction for **BST** defined in the **IEEE Std 1149.1**.
SAMPLE/PRELOAD –	Optional instruction for **BST** defined in the **IEEE Std 1149.1**.
Scan design	A technique used in digital electronics in which shift-register paths are implemented to improve testability.
Scan path	The shift-register path as used in a scan design, e.g. a **BSR** chain.
Shift	A defined operation of a serial shift register (e.g. Boundary-Scan Register) at which data is shifted through the serial path.
Short	A name of a fault caused by an electrical connection between two or more signal pins or copper tracks/wires on a **PCB**.
Signature analysis –	A technique of compressing test (result) data into a smaller number of bits, which are subsequently compared with expected (stored) data for fault detection.
SMD	Surface Mounted Devices
SR	Shift Register
Stuck-at	A name for a fault that causes a **PCB** signal connection to be fixed at logical level '1' or '0', independent of the driving signals.
STV	Sequential/Serial Test Vector, applied to a single test node over a period of time.
SVF	Serial Vector Format
System logic	As opposed to the **test logic**, the functional logic performing the required non-test functions.
TAB	Tape Automated Bonding; technology to automatically place components on a **PCB**.
TAP	Test Access Port defined by **IEEE Std 1149.1**

TCK   Test ClocK input pin in the **TAP**
TDI   Test Data Input pin in the **TAP**
TDO   Test Data Output pin in the **TAP**
Test logic   As opposed to the **core logic**, the logic performing the required test
             functions, for example a Boundary-Scan Test.
TIM   Testability IMprover; **BSR** compiler of Philips Electronics N.V.
TMS   Test Mode Select input pin in the **TAP**
TPG   Test Pattern Generator
TRST*   Test ReSeT input pin (optional) in the **TAP**
Update   A defined operation of a serial shift register element (e.g. Boundary-
             Scan Cell) at which the content of the scan cell is transferred to an
             output element.
USERCODE   User identity code, an optional instruction for **BST** defined in the
             IEEE Std 1149.1.
UUT   Unit Under Test
VHDL   **VHSIC** Hardware Description Language
VHSIC   Very High Speed Integrated Circuit
Virtual Access –
             As opposed to physical access, a term used to indicate that Boundary-
             Scan Cells can be used to access board connections and through these
             connections the inputs and outputs of non-**BST** devices (clusters).
VLSI   Very Large Scale Integration
Walking Sequence –
             A sequence of test patterns at which the test vectors contain only one
             '1' between all zeros (walking 1) or vice versa (walking '0'). This
             vector shifts though the scan path such that the logical value, either
             the '1' or the '0', is applied to all nets, one at a time.
WSI   Wafer Scale Integration; technology used to integrate electronic
             products on a wafer, instead of on a PCB for example.

REFERENCES

[1] IEEE Std 1149.1-1990: IEEE Standard Test Access Port and Boundary-Scan Architecture. Published by the Institute of Electrical and Electronics Engineers, Inc., 345 East 47th Street, New York, NY 10017, USA.

[2] F. de Jong et al.: Boundary-Scan Test, Test Methodology and Fault Modelling. Journal of Electronic Testing - Theory and Applications, Vol. 2, No. 1 March 1991, pp. 77-88.

[3] Joint Electron Device Engineering Council: Standard Manufacturer's Identification Code. JEDEC publication 106-A from JEDEC, 2001 Eye Street NW, Washington DC 20006, USA.

[4] Proposed instruction as supplement to IEEE Std 1149.1: Draft P1149.1a/D9, June 26 1992.

[5] K.D. Wagner and T.W. Williams: Enhancing Board Functional Self-Test by Concurrent Sampling. Proc. ITC 1991, pp. 633-640.

[6] N. Jarwala and C.W. Yau: Achieving Board-Level BIST Using the Boundary-Scan Master. Proc. ITC 1991, pp. 649-658.

[7] P. Raghavachari: Circuit Pack BIST from System to Factory - the MCERT Chip. Proc. ITC 1991, pp.641-648.

[8] IEEE Std. P1149.5: Standard Backplane Module Test and Maintenance (MTM) Bus Protocol.

[9] D. Bhavsar: An Architecture for Extending the IEEE Standard 1149.1 Test Access Port to System Backplanes. Proc. ITC 1991, pp.768-776.

[10] P.A. Wyatt and J.I. Raffel: Restructable VLSI - a Demonstrated Wafer Scale Technology. Proc. Intl. Conf. on WSI, 1989, pp. 13-20.

[11] D.L. Landis and P. Singh: Optimal Placement of IEEE 1149.1 Test Port and Boundary-Scan Resources for Wafer Scale Integration. Proc. ITC 1990, pp. 120-126.

[12] K.E. Posse: A Design-for-Testability Architecture for Multichip Modules. Proc. ITC 1991, pp. 113-121.

[13] W.C. Bruce et al.: Implementing 1149.1 on CMOS Microprocessors. Proc. ITC 1991, pp.879-886.

[14] L. Whetsel, Event Qualification, a Gateway to At-Speed System Testing. Proc. ITC 1990, pp. 135-141.

[15] L. Whetsel: An IEEE 1149.1 Based Logic/Signature Analyzer in a Chip. Proc. ITC 1991, pp. 869-878.

219

[16] S.P. Moreley and R.A. Marlett: Selectable Length Partial Scan, a Method to Reduce Vector Length. Proc. ITC 1991, pp. 385-392.

[17] B.I. Dervisoglu and G.E. Stong: DFT, Using Scanpath Techniques for Path-Delay Test and Measurement. Proc. ITC 1991, pp. 365-374.

[18] K.P. Parker and S. Oresjo: A Language for Describing Boundary-Scan Devices. Proc. ITC 1990, pp. 222-234.

[19] K.P. Parker and S. Oresjo: A Language for Describing Boundary-Scan Devices. Journal of Electronic Testing - Theory and Applications, Vol. 2, No. 1 March 1991, pp. 43-74.

[20] IEEE Standard Number 1076-1987: IEEE Standard VHDL Language Reference.

[21] D. Chiles and J. DeJaco: Using Boundary Scan Description Language in Design. Proc. ITC 1991, pp.865-868.

[22] M. Muris: Integrating Boundary Scan Test Into an ASIC Design Flow. Proc. ITC 1990, pp. 472-477.

[23] EDIF Version 2 0 0: EIA Interim Standard No. 44, May 1987.

[24] R.G. Bennets: Design of Testable Logic Circuits. Addison-Wesley, 1984.

[25] Software package from the alliance FLUKE and Philips Industrial Electronics B.V., Eindhoven, The Netherlands.

[26] W.F. Driessen and F. de Jong: Test Interface Layer, an Appllication Layer for Boundary-Scan testers. Paper prepared for publication.

[27] F. de Jong: Boundary-Scan Test Used at Board Level. Proc. ITC 1990, pp. 235-242.

[28] P.E. Fleming, D. McClean and P. Hansen: A Format for Transportable and Reusable Scan-Based Vectors. Texas Instruments Incorporated, October 1991.

[29] F. de Jong: Testing the Integrity of the Boundary-Scan Test Infrastructure. Proc. ITC 1991, pp. 106-112.

[30] W.K. Kautz: Testing of Faults in Wiring Interconnects. IEEE Transactions on Computers, Vol. C-23, No. 4, April 1987, pp. 358-363.

[31] A. Hassan et al.: Testing and Diagnosis of Interconnects Using Boundary-Scan Architecture. Proc. ITC 1988, pp. 126-137.

[32] P. Wagner: Interconnect Testing With Boundary-Scan. Proc. ITC 1987, pp. 52-57.

[33] N. Jarwala and C.W. Yau: A New Framework for Analyzing Test Generation and Dignosis Algorithms for Wiring Interconnects. Proc. ITC 1989, pp.63-70.

[34] C.W. Yau and N. Jarwala: A Unified Theory for Designing Optimal Test Generation and Diagnosis Algorithms for Board Interconnects. Proc. ITC 1989, pp. 71-77.

[35] J-C. Lien and M.A. Breuer: Maximal Diagnosis for Wiring Networks. Proc. ITC 1991, pp. 96-105.

[36] M.L. Flichtenbaum and G.D. Robinson: Scan Test Architectures for Digital Board Testers. Proc. ITC 1990, pp. 304-310.

[37] G.D. Robinson and J.G. Deshayes: Interconnect Testing of Boards With Partial Boundary-Scan. Proc. ITC 1990, pp. 572-581.

[38] F. de Jong and A.J. de Wind van Wijngaarden: Memory Interconnection Test at Board Level. Paper submitted to the IEEE ITC Committee, June 1992.

[39] D. Bhavsar: Testing Interconnections to Static RAMs. IEEE Design & Test of Computers, June 1991, pp. 63-71.

[40] W-T Cheng et al.: Diagnosis for Wiring Interconnects. Proc. ITC 1990, pp. 565-571.

[41] BTPG(TM), Boundary-scan Test Pattern Generator, is a Trademark of FLUKE and Philips Industrial Electronics B.V., Eindhoven, The Netherlands.

[42] P. Hansen: Testing Conventional Logic and Memory Clusters Using Boundary-Scan Devices as Virtual ATE Channels. Proc. ITC 1989, pp. 166-173.

[43] F.P.M. Beenker: Systematic and Structured Methods for Digital Board Testing. Proc. ITC 1985, pp. 380-385.

[44] C.E. Stroud: Distractions in Design For Testability and Built-In Self-Test. Proc. ITC 1991, p.1112.

[45] W. Maly: Improving the Quality of Test Education. Proc. ITC 1991, p.1119.

[46] M.A. Breuer: Obstacles and an Approach Towards Concurrent Engineering. ITC 1990, pp. 260-261.

[47] P.C. Maxwell: The Interaction of Test and Quality. Proc. ITC 1991, p. 1120.

[48] K. Rose: Quality in Test Education? Proc. ITC 1991, p.1121.

[49] C.H. House and R.L. Price: The Return Map, Tracking Product Teams. Harvard Business Review, January-February 1991.

[50] M.E. Levit and J.A. Abraham: The Economics of Scan Design. Proc. ITC 1989, pp. 869-874.

[51]   M.G. Gallup et al.:Testability Features of the 68040. Proc. ITC 1990, pp.749-757.

[52]   J. Miles et al.: A Test Economics Model & Cost-Benefit Analysis of Boundary-Scan. Proc. European Test Conference 1991, pp. 375-384.

INDEX